Sustainable Development

Carlo Aall

Sustainable Development

An Account of Why Norway only Nearly Succeeded in Creating a Sustainable Climate Transformation

Springer

Carlo Aall
Western Norway Research Institute
Sogndal, Norway

Western Norway University of Applied Sciences
Sogndal, Norway

ISBN 978-3-032-35929-2 ISBN 978-3-032-35930-8 (eBook)
https://doi.org/10.1007/978-3-032-35930-8

The original submitted manuscript has been translated into English. The translation was done using artificial intelligence. A subsequent revision was performed by the author(s) to further refine the work and to ensure that the translation is appropriate concerning content and scientific correctness. It may, however, read stylistically different from a conventional translation.

Cover illustration: The cover photograph was taken on Dyrhaugsryggen in the Hurrungane mountain range, within Jotunheimen National Park. Established in 1992, Tinderittet was a national ski mountaineering race. Starting at Nordre Dyrhaugstind (approx. 2100 metres above sea level), the course descended along Dyrhaugsryggen to the finish line at Turtagrø Hotel (870 m.a.s.l.). In 2011, after three consecutive cancellations due to a lack of snow, the organizers permanently discontinued the event—making Tinderittet widely regarded as the first Norwegian ski event to be permanently discontinued because of declining snow conditions associated with climate change. The photograph was taken on April 30, 2011, capturing three would-be participants: Åsmund Kongsvik Aall, Gunhild Sofie Vangsnes, and Jenny Johnsen Kvaale. Photo: Carlo Aall.

This Springer imprint is published by the registered company Springer Nature Switzerland AG
The registered company address is: Gewerbestrasse 11, 6330 Cham, Switzerland

If disposing of this product, please recycle the paper.

Dedicated to the memory of Karl Georg Høyer—a mentor, a colleague, and a dear friend. His pioneering work and wisdom continue to inspire.

Foreword: Reflections After a Lifetime on the Frontlines

Like Carlo Aall, I have spent a lifetime trying to create a better and more sustainable world for ordinary people. My work has primarily been international—comfortably elevated above day-to-day (short-term) politics within national borders. Meanwhile, Carlo has fought the battle where reluctance and resistance to making meaningful change for future generations is most intense: at the national and local levels.

After decades of campaigning for the well-being of future generations, I have reached a clear understanding of the primary challenge—and more importantly, the obvious solution. The defining challenges of the coming decades are (a) climate change and (b) increasing inequality. The solutions are (a) CO_2-free energy and (b) redistribution of wealth from the super-rich to collective climate solutions. The problem is that neither the issue itself, nor certainly the solutions, find broad support among the populations of wealthy market democracies.

Let me address the climate issue first. To stop the continuing rise in the global average temperature, humanity must reduce human-caused CO_2 emissions to zero. If we do not cut to zero, temperatures will continue to rise, albeit at a slightly slower pace. This transition to net zero must be implemented uncomfortably quickly—within 50 years—if we are to avoid temperatures so high that they reduce the quality of life of large portions of the world's population, regardless of what steps we take in other sectors.

Since the vast majority of human-caused CO_2 comes from burning coal, oil, and gas, there is no alternative to reducing fossil fuel use to near zero within the next fifty years, and certainly before 2100. Fortunately, it is technically possible to achieve this. It requires a transition from coal, oil, and gas to CO_2-free energy sources, such as solar panels, wind turbines, hydropower, nuclear power, and biomass. If we also deploy carbon capture and storage and direct air capture of CO_2 from the atmosphere, it is possible to reach net zero somewhat faster.

Calculations show that completing the transition in 50 years would require us to increase global spending on energy from the current 6% of gross domestic product (GDP) to 8% during these 50 years. This includes both CAPEX (capital expenditures), such as the heavy upfront costs of building wind farms and solar arrays, and OPEX (operating expenditures), the ongoing costs of running and maintaining the energy system. The challenge lies in the fact that while CO_2-free energy often has lower OPEX, the initial CAPEX required is significantly higher than for traditional fossil fuels. As a result, the return on investment in CO_2-free energy is lower than on investment in fossil energy—using standard market valuation methods. Hence, the transition to CO_2-free energy will not occur at scale unless subsidized—to the tune of $8\%–6\% = 2\%$ of GDP.

What does an increase of 2% of GDP entail? It involves shifting 2% of the world's workforce and capital away from current modes of production to CO_2-free energy. Interestingly, this cost increase is roughly equal to the increase in defense spending that European governments are now considering in response to pressure from Putin and Trump to secure another intangible value: safety and control over our own future. Thus, such an increase is politically possible, although it remains slow and difficult to achieve.

In sum, it is both technically and economically possible to create a better climate and enhanced quality of life for future generations. However, it requires that the world henceforth use a larger portion of its workforce to produce wind turbines and solar panels instead of consumer goods. The result will be (slightly) lower consumption but a (much) higher quality of life. If we can arrange for the rich to foot the bill—by reducing their rate of accumulation of wealth through increased taxes on wealth and inheritance—we will simultaneously help solve the problem of rapidly increasing wealth inequality.

To achieve the desired shift to CO_2-free energy in a capitalist market economy, such energy must be subsidized—the transition will not happen on its own. These subsidies must be around 2% of GDP for a long time. In Norway, this would mean spending around 60 billion NOK annually—roughly double the current wealth tax. The money would be used to transform the petroleum industry away from the production of profitable oil and gas toward lower-return investments such as wind power and green hydrogen for the European market.

The problem is, as mentioned, that the transformation will not occur at sufficient speed without subsidies—which require higher taxes. Understandably, most people—even in wealthy countries like Norway—resist higher taxes. There is widespread reluctance to pay more today for an uncertain, intangible benefit for grandchildren fifty years hence. Furthermore, the general aversion to increased public spending seems to increase as people become wealthier. There seems to be a tipping point beyond which affluent groups increasingly prefer to access private services (health, education, transport) rather than pay the taxes necessary for public services for all. Norway appears to be approaching this point.

The only real bright spot, as I see it, is China. Here, it is almost ironic to observe that Carlo and I have arrived at almost diametrically opposed conclusions regarding the best path—and where we advise others to seek solutions—despite sharing the same fundamental understanding of the crisis facing both nature and society. On the one hand, Carlo recommends doubling down on democracy and argues—albeit with some reservations—that the world can learn by "looking to Norway" and "looking to local governance." On the other hand, I have come to believe that the world has the most to learn by looking to a superpower that—as it happens—differs markedly from the Western ideal of liberal democratic rule.

I observe that the Chinese leadership is capable of allocating significant labor and capital to build the infrastructure they believe is needed to improve the situation for the majority, even if the investment is not "profitable" in a conventional sense. Their economy is sufficiently large that they can allocate an extra 2% of their workforce on a long-term basis to the production of solar panels, wind turbines, batteries, and electric cars. These goods will ultimately be exported at cheap prices, replacing fossil capacity in the rest of the world. I believe this will be done at sufficient scale to significantly reduce global CO_2 emissions: witness how China is already supplying cheap electric vehicles to the world. The cost to China is somewhat lower growth in consumption for its citizens, but since the economy is still growing very rapidly, I believe the current policy will be sustainable until the job is done.

So this is where we stand—Carlo and I. I have spent most of my life trying to persuade people, especially in rich countries, to use more of society's resources on a global shift to CO_2-free energy. Carlo has spent most of his life fighting for a greener and more just Norway. I am not sure he has succeeded any more than I have.

Carlo has summarized his experiences in the book you are now holding. It is an entertaining and depressing saga of how difficult it is to "create a better future for most people," even in super-rich Norway—a country swimming in CO_2-free energy and possessing more collective financial resources than perhaps any other small, democratic society in history. Although Carlo's book warns against too narrow a focus on the climate issue alone and points to many other worthy causes, I am glad he does not entirely dismiss my single-minded dream: taxing the rich to build CO_2-free energy infrastructure. I believe it is the only realistic path to the sustainable society we both work for—unless we leave the matter to the Chinese.

It is vital that we find a way to frame this core idea in a way that appeals to the majority, even if it is never supported by the super-rich.

Core reading supporting my arguments:

Randers, J. (2008). Global collapse—Fact or fiction? *Futures, 40*(10), 853–864.
Randers, J. (2012). *2052: A global forecast for the next forty years.* Chelsea Green Publishing.
Randers, J., & Maxton, G. (2016). *Ein Prozent ist genug.* Oekom Verlag. (English edition: Maxton, G., & Randers, J. (2016). *Reinventing prosperity.* Greystone Books.)
Randers, J., Rockström, J., Stoknes, P. E., Golüke, U., Collste, D., & Cornell, S. (2018). *Transformation is feasible!* Stockholm Resilience Centre.
Dixson-Declève, S., Gaffney, O., Ghosh, J., Randers, J., Rockström, J., & Stoknes, P. E. (2022). *Earth for all.* New Society Publishers.
Randers, J., & Kellerhoff, T. (2024). *Tax the rich.* Oekom Verlag.

Chandran, N., Randers, J., Zhou, J., & Dubee, F. C. (2025). *Understanding China*. Routledge.

Oslo, Norway Jørgen Randers

Preface: Why Norway Almost Succeeded

In 1987, the World Commission on Environment and Development, led by Norwegian Prime Minister Gro Harlem Brundtland, launched the global goal of sustainable development. Throughout the process leading up to the launch of the United Nations (UN) report, expectations were high—both in Norway and internationally—that Norway would take on the role of a *frontrunner* in the pursuit of sustainable development. Many factors supported these expectations:

- **A democratic country**: Norway was an early pioneer in establishing the foundations of a true liberal democracy, adopting a very radical constitution for its time (1814), which, among other things, abolished the nobility. According to the "Deliberative Democracy Index," Norway has ranked among the top ten countries since 1900; in recent years, it has consistently been among the top four, alongside countries like Sweden, Denmark, and Germany.
- **An egalitarian country**: Norway evolved from a society marked by significant economic disparities in the nineteenth century to becoming, by the 1970s, the country with the smallest income differences according to rankings based on the Gini index.
- **A peace-loving country**: Approximately 800 years have passed since Norway attacked another country in a conventional war on its own. The last instance was a naval attack on Scotland in 1263.
- **A cradle of ideas concerning the need for deep societal transformation in response to ecological crises**: Norway has fostered a long line of influential thinkers and intellectual communities that have advanced ideas about the need for radical societal shifts. These include Arne Næss and Sigmund Kvaløy, who established the eco-philosophy tradition in the early 1970s; Jørgen Randers, co-author of the seminal study *The Limits to Growth* in 1972; and Erik Dammann, who in 1974 founded a Norwegian precursor to today's degrowth movement, the organization The Future in Our Hands, which remains active today.
- **A frontrunner in the institutionalization of environmental policy**: Norway was the first country to appoint its own Minister of the Environment and establish a dedicated Ministry of the Environment (in 1972). It was also the first to secure a

minimum institutional capacity for environmental protection in local government through state funding of environmental coordinator positions in all Norwegian municipalities between 1990 and 1997, and one of the first countries to introduce a national CO_2 tax (in 1991).

- **The birthplace of the goal of sustainable development**: Norway's then-Prime Minister, Gro Harlem Brundtland, was chosen to lead the UN Commission on Environment and Development, which in 1987 presented the report often referred to as the Brundtland Report, in which the goal of sustainable development was formally articulated.
- **A long-standing foundation of renewable energy**: Long before climate change topped the political agenda, Norway developed vast hydroelectric resources, which has resulted in one of the world's highest shares of renewable energy, second only to Iceland. This energy has been available at comparatively low prices, helping place Norway at the global forefront in the adoption of electric private cars. Today, this historical legacy provides Norway with a unique, near-carbon-free baseline for its broader climate transition, largely inherited from pre-climate policy eras. With such a broad and deep foundation for societal transformation, it was reasonable that expectations for Norway as a *frontrunner* in climate and sustainability were set high.

The same year that the World Commission's report was presented, I submitted my master's thesis at the Norwegian University of Agriculture, titled "Eco-Municipality." My research examined how rural Norwegian municipalities could transition toward sustainability. By 1988, I was working as an environmental coordinator in the rural, self-declared "eco-municipality" of Ølen. In 1990, I moved to another self-declared eco-municipality—Sogndal—to begin my career as a researcher at the Western Norway Research Institute (WNRI). I have been a researcher ever since.

Sogndal is a rural town with 10,000 inhabitants. The nearest large city by Norwegian standards—Bergen—is four hours away by car and half-hour by plane. Add one more half-hour, and you reach Norway's capital, Oslo. Sogndal is characterized by its large student population and its role as one of the four campuses of the Western Norway University of Applied Sciences. WNRI, located on campus, is a small *independent research institute* with 30 employees.

In research applications to the Research Council of Norway or the EU, we often promote our region as a "research laboratory." Sogndal sits at the foot of the Jostedalsbreen glacier, the largest on the European mainland. Here we witness one of the direct consequences of climate change as the glacier shrinks year by year, affecting tourism and other sectors. Meanwhile, just below my home in Leikanger, the world's currently northernmost vineyard was established a few years ago, now producing a fresh white wine from Solaris grapes—a literal taste of a changing climate.

The Sognefjord, with Sogndal situated on its shores was ranked by *National Geographic Traveler* in 2004, together with other Western Norwegian fjords, as the world's most beautiful and sustainable travel destination. The expert panel summarized the ranking as follows: "This place has wonderful, living traditional culture, wonderful landscape, not crowded... Excellent environmental quality, local people involved in a very smooth way." Twenty years later, the Norwegian fjords still top the ranings, but now with accompanying warnings of increasing vulnerabilities. The magazine highlights new threats that have emerged since the original ranking, including overtourism, waste dumping, and "monster" power lines that scar the landscape.

The observations made by *National Geographic Traveler* capture the essence of the story I want to tell: the story of how Norway—the country with perhaps the best prerequisites for achieving a sustainable climate transformation at the dawn of what we can rightly call the "sustainability era"—fell just short of success. What are the success stories that brought us close, and what are the mistakes that explain why, 35 years later, we must admit that Norway failed to meet the challenge? And finally, what can Norway and other countries learn from this story to apply over the next 35 years? In that period, Norway and the rest of the world will have no choice but to achieve a sustainable climate transition. The consequences of continued failure will be intolerable.

Behind this narrative lies another account I wish to convey: the experience of being a researcher who wants not merely to research a subject, but to conduct research for a cause. I want to tell the story of conducting critical research, where "critical" is understood as prioritizing research on the most decisive factors for achieving sustainable development while maintaining a critical stance toward established knowledge. This approach involves challenging assumptions such as the belief that technological measures alone can solve sustainability and climate problems, and that perpetual economic growth is therefore possible. Part of this story also concerns the opportunities found in working within a small, flexible research institution located outside the major research hubs of large cities.

WNRI is arguably the research community in Norway, and perhaps internationally, that has worked the longest and most systematically on critical research in promoting the goal of sustainability. The institute's targeted sustainability research began as early as 1988, one year after the World Commission's report, with a comprehensive program called Nature- and Environment-Friendly Urban Development, supplemented two years later by the Eco-Municipality program focusing on rural areas. Sustainability research at WNRI has continued actively ever since, maintaining its critical orientation despite changing themes, and will likely continue in a similar spirit for the next 35 years. Having been a key player in this research and in the often-heated public debates it has sparked, I see myself as a *contemporary witness* to the Norwegian sustainability struggle. Specifically, I will describe the research I have conducted alongside colleagues across six areas:

- Fundamental and conceptual research on the interpretation and implementation of policies for sustainable development.
- The establishment of local environmental governance, where municipalities often acted as frontrunners, at times adopting approaches that stood in direct conflict with national policies.

- The rise and eventual stagnation of efforts to implement policies for sustainable consumption—a field that struggled to maintain political momentum and largely faded from view.
- The transformation of Norwegian tourism, tracing how major industry players were gradually compelled to accept a more radical form of responsibility, including responsibility for emissions from international travel to and from Norway.
- The development of local climate initiatives, highlighting municipalities as pioneer laboratories that often outpaced national climate ambitions.
- The emergence of climate change adaptation, a policy area in which local-level initiatives preceded and challenged the development of national frameworks.

This book is both a *critical inquiry* and an intentionally personal account. It is my story, enriched with anecdotes and experiences drawn from interactions with stakeholders and participation in public debates. To stay as close as possible to the "reality of the debate," I have made a deliberate methodological choice: I refer almost exclusively to original project notes and reports rather than academically "polished" journal articles. These reports—mostly in Norwegian, but often with English summaries—represent the raw, unedited frontline of the sustainability battle. In an age of automated translation, these primary sources are now accessible to any reader who wishes to dive into the archives of WNRI (www.vestforsk.no).

Sogndal, Norway Carlo Aall

Acknowledgements The writing of this book has been a long-term personal project, largely conducted during my own private time. However, I am deeply grateful to the Western Norway Research Institute (WNRI) for providing institutional support and partial funding toward the finalization of the manuscript.

The decades of empirical research and projects underpinning the analysis in this book have been made possible through various funding sources over the years. I would especially like to acknowledge the Research Council of Norway for its long-standing support through numerous research programs, as well as the various local, regional, and national authorities and stakeholders who have participated in and funded our sustainability and climate adaptation projects.

Finally, my warmest thanks go to my colleagues at WNRI and the Western Norway University of Applied Sciences for decades of stimulating discussions, shared insights, and dedication to critical sustainability research.

Competing Interests The author declares that the writing of this book was conducted independently. The partial funding received from WNRI carried no editorial restrictions or influence over the content. The author has no financial, personal, or professional competing interests that could be perceived as inappropriately influencing the research, arguments, or conclusions presented in this volume.

Contents

1 **The Dispute Over the True Meaning of Sustainability—The Path from Governing Power to Dilution** 1
The Political Background to the Goal of Sustainable Development 2
Norway Was Early in Developing a Sustainability Policy 5
How to Understand the Goal of Sustainable Development 7
From Sustainability Goals to Sustainability Indicators 13
The Corruption of the Political Understanding of the Sustainability Goal 17
A Shift from "Sustainable" to "Sustained" Development Indicators, and from "Political Governance" to "Administrative Reporting" 19
Where Does the Sustainability Debate Stand Today, Internationally and in Norway? 26
References 30

2 **Municipal Environmental Policy—From Government-Initiated Reform to Local Counter-Power** 33
Introduction 33
An Environmental Policy Backdrop: The Pre-institutional Period 35
Initial Attempts to Strengthen Structures for Local Environmental Policy 36
Preparations for the Major Reform Initiative Begin 37
The First Two Research Projects Aimed at Examining Municipalities' Room for Manoeuvre in Their Efforts Toward Sustainable Development 41
From State-Initiated Program to State Reform 48
Local Agenda 21 Arrives in Norway—Eventually! 50
Where Do Norwegian Municipalities Stand Today in Their Efforts to Promote Sustainable Development? 53
Deinstitutionalization of Municipal Environmental Protection Efforts 54

The Dilution of the Sustainability Goal Also Occurs at the Local Level ... 55
Specialization of Local Sustainability Efforts ... 56
References ... 57

3 Up Like a Lion and Down Like a Lamb—The Legacy of Minister of the Environment Torbjørn Berntsen in the Debate on Sustainable Consumption ... 61
Introduction ... 61
Environmental Policy Background ... 62
A Study of Barriers to the Transition to Sustainable Consumption ... 66
Oslo as the First City in the World to Undertake a Scientific Analysis of Its Residents' Ecological Footprint ... 70
The Brundtland Report Celebrates 20 Years ... 72
Experiences and the Way Forward ... 77
References ... 80

4 Sustainable Tourism—From "Best Games Ever" to "Naive Technological Optimism" ... 83
Introduction ... 83
Emergence of a New Field of Research on Sustainable Tourism and the Launch of the Slogan "Travel Less—Live More!" ... 85
Sustainable Tourism Becomes Part of Norwegian Tourism Policy ... 89
National Tourism Strategy from 2006 ... 90
Ten Principles for Sustainable Tourism—Without Reference to "Climate" ... 91
The World's Most Advanced System for Sustainable Tourism Certification ... 93
Emissions from Travel to and from Norway Are Finally Accounted for by the Tourism Industry ... 96
Cruise Becomes the Climate Scapegoat ... 98
The Struggle Over the Tourism Industry's Dependence on Air Travel ... 101
Research in Support of Sustainable and Climate-Friendly Tourism ... 104
Scenarios for Sustainable Tourism ... 105
Food and Tourism: A Success Story ... 109
Climate Change Adaptation on the Agenda in Norwegian Tourism ... 113
Why Did We Not Succeed in Creating Sustainable Tourism in Norway? ... 116
How Bad Is It for Norwegian Tourism? ... 116
Norway Has Reached Its Goal—According to Some ... 118
According to Others, We Have Not Reached the Goal—Why Not? ... 120
Hope Shattered: On the COVID-19 Pandemic as a "Perfect Storm" ... 121

Norwegian Technological Optimism ... 122
Norway, the Oil Nation ... 124
How Can a Genuine Transition to Sustainable Tourism Be Achieved? ... 125
References ... 126

5 Norway's Climate Schism—The National Oil-Driven Policy Versus Local Consumer Action ... 131
The Scientific and Political Context ... 131
Municipalities Enter the Arena of Climate Policy ... 135
Western Norway Research Institute Enters the Climate Policy Arena ... 137
From Experimentation to Normalization of the Municipalities' Role in National Climate Policy ... 139
Saving Electricity and Salvaging the Climate—By Accident ... 143
The Emergence of a Parallel Track to the State-Led Normalization of Local Climate Policy ... 145
HOPE for the Climate: Can Strategic Gaming Trigger Household Change? ... 148
Is it Also Possible to Game your Way to an Alternative Local Climate Policy? ... 151
When the State Fails: The Municipal Capacity for Climate Action ... 154
References ... 156

6 Climate Change Adaptation as Failure or a Catalyst for Critical Climate Action? ... 159
Introduction ... 159
Climate Change Adaptation Enters the Climate Policy Agenda in Norway from the Sidelines ... 161
Phase 1: Ignorance ... 161
Phase 2: Acceptance—The Sogn Og Fjordane Rebellion ... 162
Phase 3: Conditional Action—The Fear of Distraction ... 164
Phase 4: Adaptation to Natural Hazard Events ... 165
Phase 5: From Natural Hazards to Climate Risk ... 166
Research "for" Climate Change Adaptation ... 167
The Struggle for a Scientific Agenda ... 167
Flora: Implementing Norway's First Climate Change Adaptation Measure ... 169
Fredrikstad: The First Integrated Climate Plan in Norway ... 170
The World's First Scenario Model for Local Climate Vulnerability ... 171
From Climate Vulnerability Analysis to Climate Change Adaptation Strategies ... 172
The Reality Gap: Infrastructure Expansion Versus Maintenance ... 176

Beyond the Pioneers: Scaling Integration in Sogn Og Fjordane by Means of "Dugnad" ... 178
From Action Research to Critical Analysis: Shifting Focus from "How-To" to "Why-Not?" ... 180
A Theory on How the Climate and Emergency Management Sectors Can Co-create Climate Change Adaptation ... 181
Are Natural Hazard Events Caused by Severe Weather or Poor Planning? ... 181
The Battle of the Spruce: When Climate Change Mitigation Becomes a Risk ... 184
Western Norway Research Institute's Homemade Vulnerability Model Meets the IPCC Risk Framework ... 187
The Unthinkable Idea: Reducing Mobility to Mitigate Climate Risk ... 189
The Blind Spot: When the Solution Becomes the Risk ... 193
Establishment of the World's First Research Centre for Sustainable Climate Change Adaptation (Noradapt) ... 195
An Academic Community of Practice: The Birth of Noradapt ... 195
Bridging the Indicator Gap: How Noradapt Succeeded where the State Gave up ... 197
Beyond Borders: National Security and Transboundary Climate Risks ... 202
Integrating Research on the Climate and Nature Crises ... 205
The Status of Climate Change Adaptation Efforts in Norway ... 208
References ... 215

7 Conclusions and Recommendations on the Path Toward a Truly Sustainable Climate Transformation ... 221
"Look to Norway?" ... 221
Norway Misses Its Own Climate Targets Twice Over ... 222
And Behind It Lurks the Oil ... 226
The Broad Norwegian Political Support for Continuing the Fossil Age ... 231
Is Hope Found in the Municipalities? ... 232
How to Get Norwegian Environmental Policy Back on a Sustainable Track ... 237
What, Then, Can the World Learn from "Looking to Norway"? ... 239
References ... 241

List of Figures

Fig. 1.1 National CO_2 emissions under three different system boundaries for Norway, Russia, the United States, and China in 2020 24
Fig. 1.2 Annual area of natural land and agricultural soil converted per capita 25
Fig. 3.1 Total direct and indirect energy use triggered by different categories of Norwegians' leisure consumption in 2001, divided into mobile and stationary energy use. See Hille et al. (2007) 75
Fig. 4.1 Framework for describing the sustainability challenges of tourism, indicating the dominant understanding of sustainable tourism in the 1990s (intensity × stationary) and highlighting the need for an expanded understanding 99
Fig. 4.2 Dual nature of tourism's climate impact: an intensity-based perspective (emissions per activity) versus a volume-based perspective (total global emissions per activity). Data derived from the following sources: United Nations World Tourism Organization (UNWTO) (2016/2019) and Sun et al. (2022) 100
Fig. 4.3 Future prospects for Lofoten regarding snow and skiing conditions 115
Fig. 4.4 Emissions accounting for the Norwegian tourism industry compared with official CO_2 inventory of Norway, 2019 (Mt CO_2) 117
Fig. 4.5 An expanded framework for describing the sustainability challenges of tourism, which also includes the orientation of measures, indicating the dominant understanding of sustainable tourism in the 1990s and 2020s, and the need to broaden this understanding to a holistic perspective on sustainable tourism development 125

Fig. 5.1 The two faces of Norway: production of petroleum (oil, condensate, natural gas liquids, and gas) and national emissions of CO_2 from 1973 to 2024 132
Fig. 5.2 The six-tiered ladder of local climate policy 140
Fig. 5.3 Examples of consumption action card from the Housing and Food sectors in the HOPE household game 149
Fig. 5.4 Default relative distribution of consumption-related greenhouse gas emissions among four main and 16 subcategories of consumption in the TRANSFORM game 152
Fig. 5.5 The TRANSFORM game board. 1 = Current global warming prospect, 1a = Emissions reduced, 1b = Public support for the municipal council, 2 = Departments (teams), 3 = Policy card, 4 = Support token, 5 = Pen, 6 = "Burden" cards illustrating public support, 7 = Remaining emissions reduction potential per category, 8 = Blank policy cards, 9 = Emission categories, and 9a = Public acceptance level per sector 153
Fig. 6.1 Conceptual analytical model showing the relationship between climate change, societal change, total vulnerability, and adaptation strategies 173
Fig. 6.2 Comparison of official climate scenario presentations. The 2011 version (left) focused on mean values, while the current portal (right) emphasizes the uncertainty interval and model spread 174
Fig. 6.3 Development in climate vulnerability for physical infrastructure (2010–2025). Estimates of percentage increase measured in constant 2011 NOK for **a** total volume of infrastructure (measured by replacement value), **b** annual operation and maintenance expenditure, and **c** the accumulated maintenance backlog. Compiled by the author based on data from RIF, KOSTRA, Norsk Vann, and the National Transport Plan 177
Fig. 6.4 Development in total payouts for climate-related natural and water damage for all of Norway (constant 2024 NOK, 2008–2024). The linear trend highlights a significant increase in claims, particularly driven by "intrusion of water" and "flooding" 186
Fig. 6.5 Strategies addressing various components of the IPCC framework for analyzing physical climate risk 192
Fig. 6.6 Funds allocated to various research purposes and to cover damages from climate-related events 196

Fig. 6.7 Ranking of Norwegian municipalities in four levels of physical climate risk in 2023. Each point represents one municipality. The vertical white lines illustrate how the average impact increases from the year 2000 to 2050 and 2100 201
Fig. 6.8 Proportion of Norwegian municipalities that report being "very concerned" or "concerned" about eight types of climate change for the years 2017, 2021, and 2024 204
Fig. 6.9 Possible negative interaction effects between four related policy areas 206
Fig. 6.10 Norwegian figures for emission intensity (CO_2/GDP) and damage intensity (natural damage + water intrusion)/ GDP measured in 2023 NOK and indexed to 2008 = 100. Emission data and gross domestic product (GDP) are sourced from Statistics Norway, and insurance claims data are sourced from Finance Norway 209
Fig. 6.11 Distribution of allocations from the Research Council of Norway for climate research from 1998 to 2023 measured in constant 2023 prices 211
Fig. 6.12 Allocation of grants for local climate initiatives from the Norwegian Environment Agency related to climate change adaptation and emission reductions from 2016 to 2025, measured in constant 2023 prices 212
Fig. 6.13 Maintenance backlog for various categories of physical infrastructure in Norway for 2010, 2015, and 2025 measured in constant 2023 prices, compiled from data from NOU 2010, National Transport Plans, the Norwegian Association of Consulting Engineers, and the municipal–state reporting system (KOSTRA) 213
Fig. 7.1 Annual changes in Norwegian greenhouse gas (GHG) emissions compared with the prevailing national climate targets since 1990, where the total assumed emission reduction for each goal-period is distributed evenly per year . . . 223
Fig. 7.2 Example of a "methodological error" frequently made by environmental authorities when presenting the desired development of greenhouse gas emissions from present to a future year (figure taken from a Norwegian government report) 224
Fig. 7.3 Emission pathways that, with a 67% probability, are consistent with the 1.5 °C and 2 °C targets based on the Global Carbon Budget and assuming an equal percentage reduction for all countries 225

Fig. 7.4 Relative development of total petroleum production in Norway (2004 = 100) and annual numbers of exploration and production licenses awarded in the Barents Sea, North Sea, and Norwegian Sea 227
Fig. 7.5 Indicator for the development of political influence for the parties that the environmental movement in Norway ranks as most concerned with climate and sustainability (the Green Party, the Liberal Party, the Socialist Left Party, and the Red Party) at the national and local levels 233

List of Tables

Table 1.1	Key characteristics of the concept of sustainable development (see Footnote 3)	11
Table 1.2	Policy areas in which I have participated in the development and testing of sustainability indicators	12
Table 1.3	Classification of core indicators and supplementary indicators in the indicator system "Sustainable Municipalities in Practice"	15
Table 1.4	First official version of national sustainability indicators for Norway presented by the Stoltenberg II government	22
Table 1.5	Two prevailing interpretations of the sustainability goal (adapted from Aall, 2021) (see Footnote 12)	27
Table 3.1	Distribution of and examples of obstacles experienced by 7 municipalities in 60 sub-projects on sustainable consumption (see Footnote 4)	69
Table 4.1	Innovation Norway's ten principles for sustainable tourism	92
Table 4.2	Criteria and indicators related to "climate" in the Norwegian system for sustainable destination certification	94
Table 4.3	Development of Norwegians' passenger transport activity 1985–2005, passenger-kilometers and percentages	102
Table 4.4	Norwegian climate measures that can be selected and scaled in the game SUSNOR	106
Table 4.5	Comparison of scenarios for development from 2005 to 2025 created using the SUSNOR game and actual development from 2005 to 2019	109
Table 4.6	Seven recommendations on how to make food services in the tourism industry more climate-friendly	114
Table 5.1	Range of greenhouse gas emissions (total, per capita, and per km^2) compared to population size, land area, and municipality type (see Footnote 22)	146

Table 6.1 Elements included in our proposal for an integrated climate policy plan 179
Table 6.2 Characteristics of existing institutions for climate policy versus civil protection that may influence the emergence of climate change adaptation as a distinct policy area 182
Table 6.3 Indicators used in the Noradapt system for risk-ranking of municipalities 200

Chapter 1
The Dispute Over the True Meaning of Sustainability—The Path from Governing Power to Dilution

Abstract This chapter analyzes the period following the launch of the Brundtland Report in 1987, when Norway had a historic opportunity to lead a fundamental transformation of societal development. The chapter demonstrates how Norwegian authorities and research communities were pioneers in identifying the radical implications of the sustainability goal: the recognition of absolute ecological limits and the necessity of reducing consumption levels in wealthy countries. The political project of constructing the goal of sustainable development aimed to unite two debates with roots in the 1960s: the environmental debate on saving ecosystems from total collapse, and the developmental debate on reducing economic disparities between so-called "industrialized" and "developing" countries. The work to develop this new goal drew on two academic discourses: the environmental discourse related to the sustainability component, and the discourse on needs related to the developmental component. However, a decisive choice was made in the political process of defining the essence of the sustainability goal. The 1970s criticism of perpetual economic growth was not only excluded; the concept was also overturned. Growth came to be seen as integral to the solution rather than a barrier to achieving sustainable development. In the Norwegian sustainability debate, this "misprogramming" was addressed in both academic and political discourse. Several attempts were made to introduce questions about the need to change the population's level of consumption—not just consumption patterns—and, more generally, the necessity for wealthy countries to accept a reduction in material consumption. However, these occasional efforts to radicalize the Norwegian sustainability debate were eventually "hijacked" by international developments in the sustainability discourse. This development happened along four axes: an expansion from an original dual concept (ecological and social sustainability) to a threefold concept (environmental, social, and economic); a lack of willingness to prioritize among the various aspects and sub-goals of sustainable development; an under-communication of the need to accept absolute environmental limits to economic development; and a continual broadening of the definition of needs included in the goal of social sustainability.

C. Aall, *Sustainable Development*, https://doi.org/10.1007/978-3-032-35930-8_1

The Political Background to the Goal of Sustainable Development

The goal of sustainable development was introduced in 1987 in the report *Our Common Future*, presented by the World Commission on Environment and Development. The United Nations established the commission in 1983 and appointed Gro Harlem Brundtland of the Norwegian Labor Party (*Arbeiderpartiet*) as its chair. Brundtland was then the Norwegian Prime Minister—the first woman to hold the office—and had previously also served as Norway's first female Minister of the Environment.

The political motive for linking "sustainability" and "development" was a desire to bridge a divide that emerged in the 1970s between so-called industrialized countries (developed countries) and developing countries. This divide became particularly evident during and after the first international environmental conference with broad high-level political participation—the United Nations Conference on the Human Environment, held from 5–16 June 1972 in Stockholm—later often referred to as the "Stockholm Conference." Developed countries wanted a unified international effort to reduce environmental problems, while developing countries believed that developed countries had created these problems and should therefore solve them alone. The developing countries, for their part, wanted developed countries to contribute to a more equitable global distribution of resources.

At a meeting of the countries of the Non-Aligned Movement in Algiers the year after the Stockholm Conference, the demand for a New International Economic Order (NIEO) was put forward. This demand included several elements, such as more stable commodity prices, increased manufacturing in developing countries, and improved market access for products from developing countries in developed countries. In 1974, the United Nations decided to initiate a series of conferences on these issues within the framework of the "intergovernmental organization within the United Nations Secretariat that promotes the interests of developing countries in world trade" (United Nations Conference on Trade and Development [UNCTAD]). In 1976, a program was adopted to address one of the demands from the 1973 Algiers meeting: stabilizing the prices of a range of commodities. However, the program was never implemented.

The main reason it did not materialize—and why the idea of the NIEO eventually faded from the international political agenda—was the oil crisis in the winter of 1973–1974 and its subsequent political aftershocks in developed countries. The immediate cause of the oil crisis was the Yom Kippur War, fought from 6 to 25 October 1973, between Egypt and Syria on one side and Israel (with substantial military support from the United States) on the other. After the hostilities ended, Arab oil-producing states imposed a politically motivated reduction in oil production and supply while simultaneously raising prices sharply. This led to a temporary shortage of gasoline, as well as a long-term shift in the global balance of power. This power shift also resulted in the near-overnight disappearance of support among developed countries for higher prices of commodities originating from developing countries.

The Four Weekends in 1973 When We Had the Streets to Ourselves for Sledging

I grew up in Tromsø, a city located as far north as Alaska's northern coast but warmed by the Gulf Stream. Summers are mild, winters aren't freezing cold, and there's plenty of snow—which we loved as kids. The winter activity that excited us most was riding sledges with steering wheels along icy roads. This was, of course, strictly forbidden, with a rare exception of four weekends in December 1973.

For four weeks, private car use was banned on weekends due to fears of impending gasoline rationing. Children everywhere in Norway loved the consequences of this car ban, and to this day, I haven't heard a single person complain. Everyone who remembers that time talks about it with a sparkle in their eyes, remembering the freedom and the rare chance for children to reclaim the streets of their neighbourhoods.

One event stands out in most Norwegians' minds: the day when Norway's regent, King Olav V (born 1903, regent from 1957 until his death in 1991), was spotted on the tram in Oslo on his way to a skiing trip. The king, of course, was exempt from the car ban but chose to leave his car behind and travel by tram, paying for his ticket just like everyone else, in solidarity with his people.

The iconic image of King Olav V on the tram in Oslo during the 1973 oil crisis. © Jan Greve, Scan-Foto/NTB.

The United Nations established the World Commission on Environment and Development to reconnect the two tracks of solving global environmental problems and achieving a more equitable distribution between the wealthy and poor parts of the world. The commission worked for four years, commissioning more than 75 expert reports and organizing a series of open conferences and hearings. These included eight public hearings held on different continents, where authorities, researchers, industry, NGOs, and the general public could provide input.

In the process leading up to the commission's report, several critical questions were raised that later shaped the ongoing debate on how to understand and implement the sustainability goal. One question concerned the role of economic growth in relation to both global environmental and developmental issues. Many critics argued that the commission failed to address the growth imperative, which they saw as highly unfortunate. Another critical question concerned nuclear power. The commission was unable to conclude whether the benefits of nuclear power—low CO_2 intensity—outweighed the significant concerns related to safety, waste, and the proliferation of nuclear weapons technology. The commission therefore presented nuclear power as an "unresolved" problem in its report. The commission also did not address environmental issues related to the military or the role of multinational corporations in environmental and developmental problems. Nor, curiously enough, did it address issues related to the world's largest industry: tourism. I will return to this last point in a separate chapter.

Nevertheless, the question of economic growth has remained central to the debate both during the commission's work and after the publication of its report. In the 1960s and 1970s, the relationship between economic growth and environmental and developmental issues was subject to considerable discussion. Several contributors to the debate pointed out that the goal of economic growth appeared to be the main driving force behind ever-increasing environmental and developmental problems. The commission concluded otherwise: economic growth is not part of the problem but rather part of the solution to environmental and developmental challenges. The commission went so far as to prescribe an annual increase in national income of approximately 5% in developing countries in Asia, 5.5% in Latin America, 6% in Africa and West Asia, and 3–4% in industrialized countries (p. 47).[1] It is important to note, however, that the commission presented growth in the wealthy part of the world as a means to achieve the sustainability goal, not as an end in itself.

In the spring of 1987, the World Commission's report was presented to the UN General Assembly. Since then, the UN has clarified and elaborated on the sustainability goal in three stages: in 1992 through the UN's action plan for sustainable development (Agenda 21); in 2000 through the Millennium Development Goals (MDGs); and in 2015 through the Sustainable Development Goals (SDGs). Since the report was presented, the number of goals has increased from a single overarching goal ("sustainable development") to 17 main goals, 169 targets, and 232 associated indicators.

Various Norwegian governments have presented a series of white papers addressing how Norway should achieve the goal of sustainable development at an overarching level (7 by 2025). They have also produced an even greater number of sector-specific white papers within climate, energy, marine management, land

[1] World Commission on Environment and Development (WCED) (1987).

use and biodiversity, agriculture, and the circular economy (18 by 2025). These sector-specific papers addressed how each area should pursue the sustainability goal.

Gro Swept Growth Criticism Under the Rug—And Thus Paved the Way for the Subsequent Corruption of the Sustainability Concept

Øystein Dahle, a former Norwegian industrialist, played a key role in developing the sustainability goal. He was once the deputy managing director of Esso Norway but later "jumped ship" to lead the Norwegian Trekking Association (DNT) and the Worldwatch Institute. Throughout this period, he also served as a personal secretary and advisor to Gro Harlem Brundtland during her tenure as chair of the World Commission on Environment and Development.

In 2016, the Western Norway University of Applied Sciences launched the country's first two-year master's programs covering climate change mitigation and adaptation, all set within the boundaries of the sustainability goal. That same year, I was appointed to a 20% position as professor of sustainable development, responsible for a master's course called "Societal Transformation." Much of this book's material will be included in the curriculum for that course. Each year, I organize a three-day introductory seminar for the course at my cabin in Valdres, which has neither electricity nor running water. For several years, I used to bring Øystein Dahle along, and as an avid outdoorsman, he enjoyed staying at such a primitive cabin. He also enjoyed the company of the students, and would often convey his main message as follows:

> If we in the wealthy part of the world are to genuinely contribute to achieving the sustainability goal, we must reduce our material consumption not *by* ten percent, but *to* ten percent.

Øystein told us he had shared this realization with Gro during their work on the World Commission's report. He said he urged her repeatedly to address it. But she was a seasoned politician, and she always replied, "If we include growth criticism, we will never get this report adopted."

So, the report was approved, but the absence of growth criticism paved the way for what critics, including myself, view as the sustainability goal's corrupted twin: "sustainable growth" and the idea that "economic" sustainability is a third dimension, on par with "social" and "ecological" sustainability.

Norway Was Early in Developing a Sustainability Policy

As might be expected, Norwegian authorities responded quickly to the World Commission's report. In the summer of 1987, two months after the report was presented to the UN General Assembly, the Norwegian Ministry of the Environment set out to draft a white paper dedicated solely to sustainable development. Exactly two years after the World Commission's report was published, the Norwegian government presented White Paper No. 46 (1988–1989), *Environment and Development. Norway's Follow-up to the World Commission on Environment and Development.*

Nevertheless, or perhaps because of the short timeframe the authorities had set for themselves, the report gave few indications that the sustainability goal had introduced anything fundamentally new to Norwegian environmental policy. Norway was the first country in the world to establish a dedicated Ministry of the Environment with an associated minister, and it had maintained a clear and well-developed environmental policy for more than 15 years. What was new, however, was the introduction of concrete and, at the time, ambitious targets in the areas of climate and energy compared to other countries (p. 58)[2]:

> Reduce the growth in CO_2 emissions so that they stabilize during the 1990s and, at the latest, by the year 2000. The government expects that emissions can subsequently be reduced.
>
> The government will... work toward a gradual leveling off of total energy consumption toward the turn of the century.

It eventually became clear that none of these targets would be achieved. Although CO_2 emissions did begin to level off in the second half of the 1990s, by the turn of the century emissions were still 24% higher than in 1989. From the turn of the century up to 2024, emissions declined somewhat; however, in 2024 they remained nearly 10% above the 1989 level. Nor did developments in the energy sector proceed as the government had hoped in 1989. Up to the turn of the century, domestic energy consumption continued to rise (by 12%), and only after the turn of the century has there been anything resembling a plateau. Nevertheless, energy use in 2024 is still 16% higher than in 1990.

An important part of the preparatory work for the white paper was distributing the World Commission's report to 600 private and public institutions, including municipalities, county authorities, political parties, and voluntary organizations. A total of 167 consultation responses were received, 25 of which came from various research institutions and universities. At the time the World Commission's report was distributed for consultation, Western Norway Research Institute (WNRI) had not yet established its environmental research group; this was also three years before I joined the institute. Nevertheless, the summary of input from the Council for Natural and Environmental Sciences at the University of Oslo is worth reading. At that time, the council had an academic orientation very similar to WNRI's later profile and that of the institute's key partner in the field of sustainability, the Programme for Research and Documentation for a Sustainable Society (ProSus) at the University of Oslo.

The council noted that the type of interdisciplinary research called for in the World Commission's report was underdeveloped in Norway. It pointed out that research across faculty boundaries was not common at Norwegian universities. The council also indicated that a lower rate of economic growth than the World Commission proposed for industrialized countries, combined with reduced energy consumption, could provide society greater freedom in choosing developmental paths. This, in turn,

---

[2] Miljøverndepartementet (1989).

could help achieve the goal of sustainable development. Furthermore, the council remarked that if Norway, with its unique conditions, was unable to meet the World Commission's requirements for lower and more efficient energy use, it would be difficult to imagine which other countries could.

It Would Take Almost 30 Years from the Launch of the World Commission's Report Before a Norwegian University Established a Position Titled "Professor of Sustainable Development"

White papers to the Norwegian Parliament primarily contain political guidelines and only rarely mention specific measures that involve the allocation of funds. This was also the case for White Paper 46 (1988–89)—*Environment and Development. Norway's Follow-up to the World Commission on Environment and Development*. One exception, however, was the following promise:

> To ensure the interdisciplinary aspect of the development now underway at the universities, positions at the professorial level will be established under the Ministry of the Environment's budget during 1989.

In the subsequent national budget, the promised allocation was granted for the establishment of no fewer than five professorships—one position at each of the then four universities in Norway (Oslo, Bergen, Trondheim, and Tromsø), in addition to the Norwegian University of Agriculture, where I studied.

Then began the dispute over which department should be awarded the professorships. Neither the four universities nor the Norwegian University of Agriculture could agree internally. After a few years of inconclusiveness, the ministry lost patience, and the allocation was withdrawn.

In 2012, I was awarded a professorial competence in sustainable development at Aarhus University. In 2016—almost 30 years after the World Commission's report was published—Western Norway University of Applied Sciences established a 20% position titled "Professor of Sustainable Development." I was awarded that position in connection with the establishment of a two-year international master's program called "Climate Change Management." To this day, this remains the only professorship at a Norwegian university is titled simply "sustainable development."

How to Understand the Goal of Sustainable Development

Sustainable development should first and foremost be understood as a political objective shaped through political processes. Hence it will be, and ought to be, subject to ongoing political debate regarding its content. The academic discourse must relate to and reflect this reality. For *this* reason, academic practice should analyze and, if necessary, critically examine how the content of the objective evolves over time, rather than reporting passively on its development. This dual role is demanding, as researchers can easily be criticized for adopting a political stance in support of a particular interpretation of the objective's content.

Over the years, researchers within the environmental research unit at WNRI—including Karl Georg Høyer, Erling Holden, Hans Jakob Walnum, and myself—have produced an extensive body of work, ranging from reports and scientific articles to doctoral dissertations. This research has focused on conceptual understandings and practical applications of sustainability within sectors such as tourism, transport, industrial development, and municipal administration.

An important part of my own professional development in this field, and that of my colleagues at WNRI, was the establishment of a long-term collaboration with Professor William M. Lafferty and ProSus—the Program for Research and Documentation for a Sustainable Society. He led the program throughout its existence, until 2009. ProSus was a research program at the University of Oslo established by the Research Council of Norway as a continuation and reorganization of an earlier initiative called the Alternative Future Inquiry.

The Alternative Future Inquiry ran from 1986 to 1996. It was initiated by 17 voluntary organizations within the fields of environment, peace, solidarity, and gender equality, and it received funding through a resolution of the Norwegian Parliament. A central initiator of the inquiry was Erik Dammann and the organization the Future in Our Hands (FIVH), which was established in 1974. The purpose of the inquiry was to explore alternatives to the dominant model of societal development, which was rooted in the notion of perpetual economic growth. ProSus continued the work of the Alternative Future Inquiry in 1996, but with a less normative mandate: to evaluate Norway's efforts toward sustainable development at both national and local levels. That mandate fit perfectly with the research profile of our institute.

Italian Patent Leather Shoes Meet Western Norwegian Hiking Boots

My collaboration with William M. Lafferty—Bill among friends—began when I knocked on his office door in Oslo and asked if he would like to hear about our then-flagship project on municipal environmental auditing. "I'd be happy to", Bill replied. He became so enthusiastic that he proposed a professional collaboration between ProSus and WNRI. We quickly agreed to formalize the partnership by appointing him as an adjunct professor at WNRI.

As the collaboration between Bill and me progressed, it seemed logical to introduce the other members of our respective research groups. We agreed to organize a social gathering, and I suggested a meeting point halfway between Oslo and Sogndal that was easily accessible by public transport.

The gathering was to take place in the spring. For those of us living in Western Norway, it was natural to suggest the Norwegian Trekking Association's (DNT) lodge in the mountain hamlet of Finse. Accessible only by the Bergen Line, Finse station sits at an elevation of 1222 m (4,009 ft) above sea level, making it the highest point in the Norwegian railway system.

Finse holds iconic status for Norwegians who love skiing, especially on spring snow. Since there are no roads, the only way to reach it is by train. When the ground is still covered by snow, the only thing to do there—apart from enjoying the sunshine—is to go skiing. In our naivety, we assumed this was common knowledge, but that was certainly not the case.

While all members of the WNRI group had equipped themselves with skis, backpacks, and sturdy mountain gear, the entire ProSus group—without exception—stepped off the train, carrying suitcases and wearing smart clothes and dress shoes. This made the 200-m trek across the crusted snow to the DNT lodge a significant challenge. The mood was not improved when, just 10 m from the platform, we had to pass Hotel Finse 1222, a traditional high-standard hotel. The less-than-ideal start culminated when poor Bill, suffering from the onset of a slipped disc that was aggravated by trudging through the snow, finally arrived at the reception desk and asked for a single room. He was told, "No. As with all of DNT's lodges, there are no single rooms here. Only four-bed rooms."

Oluf Langhelle, a ProSus member and close colleague of Bill's, whispered to me, "Don't say anything. Just wait; this will pass." Fortunately, his prediction proved true. We had a productive and academically rewarding seminar that laid the foundation for a long-lasting professional collaboration. My friendship with Bill also survived the Finse encounter, evolving into a close bond and a mutual tradition of celebrating each other's personal milestones in the many years that followed.

The focus of this joint research profile at Western Norway Research Institute and ProSus primarily consisted of three elements:

- A critical analysis of the *intentions* behind the sustainability goal and the derivation of a *fundamental understanding* of the content of the goal that is consistent with those intentions.
- Development and testing of *tools* for the *evaluation* and *management* of sustainability efforts in close dialogue with user interests across various policy areas.
- Evaluation of how the understanding and implementation of the sustainability goal *change over time* within different policy areas.

The World Commission's report itself allows for highly diverse and, in principle, contradictory interpretations of the sustainability goal. Already five years after the publication of the World Commission's report, it was possible to identify 70 different definitions of the goal in public debate.[3] At the same time, despite academic and political disagreements regarding the practical implications of a sustainability-oriented policy, we believed it was possible to identify some fundamental key characteristics

[3] Høyer (1997).

underlying the intentions behind the goal of sustainable development. This perspective, first formulated by Karl Georg Høyer in our research group in 1991,[4] was later elaborated in his doctoral dissertation and, three years later,[3] in my own dissertation. We also eventually found that the same line of thought was adopted by William M. Lafferty and his colleagues at ProSus as the basis for their work.[5]

So what are these key characteristics? Our reflections were based on the following two key quotations from the Brundtland Report.[1] First, there is the most frequently cited short definition of the concept (p. 42):

> Sustainable development is development that meets the needs of the present without compromising the ability of future generations to meet their own needs.

Then comes an addition that is crucial for our derivation of a key characterization of the sustainability goal (p. 42, my emphasis):

> [Sustainable development] encompasses two key elements: meeting needs, particularly the *basic needs* of the *world's poor*, which should be given *highest priority*; [and] the concept of *limitations* imposed by *current technology* and *social organization* on the environment's ability to meet present and future needs.

The first of the two key points is particularly important for understanding discussions related to terms such as "social" and "economic" sustainability.[6] The second is important for discussions about expectations regarding the contributions of technological development and the emphasis on technological measures in achieving sustainable development,[7] as well as discussions related to concepts such as decoupling, eco-modernization, and degrowth.[8]

The key characteristics we eventually agreed upon with the researchers at ProSus were as follows:

- Sustainable development must be fundamentally understood as a *twofold objective*, encompassing both an ecological dimension ("sustainable") and a social dimension ("development"), as reflected in the derived goals of ecological and social sustainability, respectively. However, compliance with the first element, ecological sustainability, must be regarded as a prerequisite for (and therefore imposes limitations on) the achievement of the second element, social sustainability.
- *Ecological* sustainability encompasses the principles of *nature's carrying capacity*, meaning that absolute physical ecological limits must not be exceeded; the *precautionary principle*, meaning that when there is fundamental uncertainty regarding the risk of irreversible ecological consequences from interventions in nature, such interventions must be halted; and criteria related to *time* (long-term perspective) and *space* (the relationship between global, national, and local levels) in the management of ecosystems.

[4] Høyer (1991).

[5] Lafferty and Langhelle (1999).

[6] Aall (2014).

[7] Walnum et al. (2014).

[8] Xue et al. (2017).

- *Social* sustainability includes a *developmental orientation*, meaning the prioritization of meeting basic human needs; *intergenerational justice*, meaning efforts to ensure a fair distribution of benefits and burdens over *time* between different generations; and *global justice*, meaning efforts to ensure a fair distribution of benefits and burdens across *space*, with particular emphasis on the relationship between poor and rich countries worldwide.
- *Goal hierarchy*, meaning that when the sustainability objective is divided into sub-goals, these must be ranked so that subordinate sub-goals serve as means to achieve higher-level sub-goals in the hierarchy.

An implication of this understanding, which is particularly relevant to the practical political sustainability discourse, is the recognition that economic growth must be treated as a subsidiary goal within the context of sustainability. It can therefore only be pursued as long as it does not conflict with considerations of ecological or social sustainability.

My colleague and then research leader for the environmental group at Western Norway Research Institute, Karl Georg Høyer, developed in his doctoral work a hierarchy for the operationalization of the sustainability goal into 15 sub-goals, which several of us in the environmental group found highly useful in our subsequent efforts to operationalize the sustainability goal within various policy areas (see Table 1.1).

The hierarchy presented in the table consists of three levels, using terminology derived from thermodynamics: extra prima, prima, and secunda. Extra prima refers to energy—or other natural resources—at the highest quality levels. Prima denotes a lower, yet still very high, quality. Extra prima and prima primarily correspond to what are considered the "central" characteristics of the sustainability goal. The "secunda"

Table 1.1 Key characteristics of the concept of sustainable development (see Footnote 3)

<table>
<tr><td colspan="7">Extra prima</td></tr>
<tr><td colspan="3">Satisfy basic needs</td><td colspan="4">Ecological sustainability</td></tr>
<tr><td colspan="7">Prima</td></tr>
<tr><td>Intrinsic value of nature</td><td>Long-term perspective</td><td>Equitable global distribution of benefits and burdens</td><td>Equitable distribution of benefits and burdens over time</td><td>Cause-oriented environmental protection</td><td colspan="2">Public and popular participation</td></tr>
<tr><td colspan="7">Secunda</td></tr>
<tr><td>Reduction of current total energy consumption in wealthy countries</td><td>Reduced emissions of greenhouse gases, especially CO_2</td><td>Reduction of current consumption of non-renewable energy and material resources in wealthy countries</td><td>Increase in current consumption of renewable energy and material resources</td><td>Pollution levels within the ecosystem's tolerance limits</td><td>Prioritization of technological development for efficient utilization of natural resources</td><td>Transfer of resources and technology from wealthy to poor countries</td></tr>
</table>

level is included to contextualize characteristics that are prominent in debates about sustainability indicators. The term "secunda" implies that these characteristics can be derived from extra prima and prima characteristics.

For example, reducing greenhouse gas emissions is necessary based on an overarching demand for ecological sustainability. However, not all forms of policy aimed at reducing emissions are therefore consistent with sustainable development. Equitable distribution—both globally and over time—and the prioritization of meeting basic needs are also required. The logic and structure outlined below formed the basis for our subsequent attempts to develop indicators in dialogue with various user groups (see Table 1.2).

Sustainable Development? I Don't Want That!

After many years as a "door-to-door salesman for the sustainability goal," William M. Lafferty shared an important experience with me. Early in his work, he was struck by how many people he didn't expect to be interested in the topic nevertheless contacted him and asked to hear about sustainable development. Then, after a few years, he found that many of these same people came back and

Table 1.2 Policy areas in which I have participated in the development and testing of sustainability indicators

Policy area	Indicator theme	Client
Sustainable development	Local agenda 21	Norwegian association of local and regional authorities
	Ecological footprint	Oslo municipality
Greenhouse gas emissions	Direct and indirect greenhouse gas emissions	Stavanger municipality
		Sogn og Fjordane County municipality
Adaptation to climate change	Climate vulnerability	Arctic council
	Physical climate risk	Norwegian Centre for Sustainable Climate Adaptation (Noradapt)
	Physical climate risk	Nordland County municipality
	Physical climate risk, transboundary climate risk, transition risk	Troms County municipality
Transport	Sustainable passenger transport	Oslo Sporveier
	Sustainable freight transport	Nordic transport policy network
Tourism	Sustainable tourism	Sogn og Fjordane County municipality

said, "Now I understand your explanation of the sustainability goal, and I don't want it!" He interpreted this as a sign of quality for the sustainability goal—that it had enough substance to provoke resistance, unlike those overarching goals that everyone can agree on and therefore have no practical application in steering society in a different direction.

From Sustainability Goals to Sustainability Indicators

A quote from a one of the co-authors of *The Limits to Growth—Donella Meadows* — is often cited in discussions about adopting alternatives to economic indicators such as gross national product (GNP) for measuring development[9]:

> The indicators a society chooses to report to itself about itself are surprisingly powerful. They reflect collective values and inform collective decisions. A nation that closely monitors its salmon populations or the safety of its streets makes different decisions than a nation that focuses solely on its GDP.

The most extensive area of research I have been involved in over the years is precisely the development and testing of such alternatives—known as sustainability indicators—within various policy domains. In most cases, this work has been carried out in close collaboration with those who will be using the indicators.

An important initial task in this area was reviewing research and development on environmental and sustainability indicators,[10] which provided us with the opportunity to familiarize ourselves with fundamental indicator theory and practical experience in developing and applying indicators. Particularly significant was a summary of a major international conference specifically focused on environmental and sustainability indicators, which proved highly useful for our subsequent work on indicators.[11] The authors distinguished between "cold" and "hot" indicators and emphasized the importance of establishing indicators that are "appropriately warm." Cold indicators are scientifically accurate but may be more difficult for decision-makers to understand and relate to. Hot indicators are often perceived as more relevant for decision-making but may be less scientifically precise. The challenge lies in achieving indicators and indices that are "appropriately warm."

The authors further pointed out that surprisingly often, little effort has been invested in clarifying the intended use of indicators and, consequently, in identifying the target audience for them. Different target groups, purposes, and areas of application may require different designs and orientations of indicators. In a summary

---

[9] Meadows (1993, May 20).

[10] Høyer and Aall (1997).

[11] MacGillivray and Zadek (1995).

of 30 years of experience developing and testing environmental and sustainability indicators, I distinguish between six purposes for employing indicators[12]:

- Trend analysis: Describing developmental trends in nature and society.
- Impact assessment: Describing the consequences of plans or planned actions.
- Evaluation: Recording or evaluating the effects of implemented actions.
- Monitoring: Recording and monitoring the state or development of a system.
- Reporting: Reporting upward within a decision-making system.
- Comparison: Comparing the development or state between two systems (e.g., municipalities).

Furthermore, I distinguish between three *governance-related contexts* in which indicators may be applied:

- Information and debate, targeting the general public, stakeholders, and interest groups in society.
- Political governance, targeting elected officials at various administrative levels.
- Administrative governance, targeting employees in public administration at various administrative levels.

An index or indicator system may serve multiple purposes and, consequently, address several target groups. However, different combinations of objectives and governance contexts (and thus target groups) may require different designs of indices and indicator systems. We carried these early insights forward in our work on developing and testing indicators in various policy settings.

In terms of content, the various indicator systems we developed and tested were based on the overarching typology of the sustainability goal established by WNRI and ProSus in the 1990s (see also Table 1.1). We also drew on several literature reviews of experiences with environmental and sustainability indicators[13] and indicators related to climate adaptation.[14]

The main effort has been the development of indicators within the framework of *municipal environmental management* (see also the chapter on local environmental policy), but we have also developed and tested indicator systems within the fields of transport and tourism (for the latter, see also the chapter on sustainable tourism).

The first national system I helped develop was created in the early 2000s in collaboration with the Foundation Idébanken for the Norwegian Association of Local and Regional Authorities (KS). It was linked to the KS's initiative on Local Agenda 21. The indicators were divided into seven sustainability themes, each with an associated core indicator, and municipalities were encouraged to develop their own supplementary local indicators (see Table 1.3).

The indicator system was developed in dialogue with and tested in five municipalities and districts (Porsgrunn, Gamle Oslo District, Sund, Tingvoll, Vega) and two county municipalities (Akershus, Sogn og Fjordane). After the trial period, we

[12] Aall (2021).

[13] Aall and Norland (2005).

[14] Aall and Norland (2003), Næss et al. (2006), Heiberg et al. (2008) and Rød et al. (2025a).

Table 1.3 Classification of core indicators and supplementary indicators in the indicator system "Sustainable Municipalities in Practice"[15]

Sustainability theme	Core indicator	Example of supplementary indicator
Quality of life	Living conditions index	Life expectancy
Local pollution	Emissions of phosphorus and nitrogen from households and municipal wastewater	Number of residents affected by road traffic noise, estimated using the noise annoyance index
Resource management	Amount of household waste per capita	Household waste sent for recycling
Biodiversity	Approved applications for exemptions for activities in agricultural, natural, and recreational areas (percent)	Number of planning and development cases where the population status of vulnerable or threatened species has been negatively affected
Energy and climate	Emissions of CO_2 equivalents per capita	Number of passenger cars per capita
Sustainable production and consumption	Proportion of private companies and municipal enterprises that are environmentally certified	Usable floor area per dwelling
Municipal operations	Proportion of municipal employees who have completed environmental training	Energy consumption in municipal buildings per employee

proposed that the KS establish the system digitally as a standardized service for its member municipalities. The system was available online, but only for a short period. As the momentum behind Local Agenda 21 diminished during the 2000s and municipal environmental work became more specialized in climate initiatives, the system was discontinued.

At the same time as my work for the KS, I contributed to pioneering academic work for Oslo Municipality. For the first time in Norway—and among the first internationally—we applied the "ecological footprint" methodology at a local rather than national level, which was the original scope of the method.[16] This methodology includes the indirect environmental impact of private consumption in calculating local environmental pressure. It also addresses social sustainability and global equity by referencing the globally available footprint per person. Through this report, Oslo Municipality began its role as a national leader in addressing both direct and indirect environmental issues related to private consumption—a position it has maintained despite changing political majorities.

The year before the work for Oslo Municipality began, a similar project was initiated for Stavanger Municipality. It involved developing climate indicators that

[15] Aall et al. (2002).

[16] Aall and Norland (2002).

also included indirect emissions associated with private consumption, but without converting emissions and energy use into an ecological footprint.[17] The success of this project lies in establishing Norway's first digital, web-based climate calculator for estimating direct and indirect greenhouse gas emissions, energy use, and ecological footprint resulting from private consumption.[18]

Ten years later, we elevated the footprint calculations one level up in the decision-making hierarchy. In our own county—Sogn og Fjordane—we demonstrated how to calculate the average personal ecological footprint for a region.[19]

The most extensive and long-standing effort in indicator development has nevertheless been within the policy area of climate adaptation. As early as 2005, I participated in developing an initial proposal for a set of indicators linked to the concept of climate vulnerability. The Arctic Council commissioned the work, and we developed and demonstrated a set of indicators in the municipalities of Norway's three northernmost counties: Nordland, Troms, and Finnmark.[20] In the first project, we tested a typology for three categories of indicators—natural vulnerability, institutional vulnerability, and socioeconomic vulnerability—developed in collaboration with researchers from, among others, ProSus.[14] Our three categories of indicators would later prove to align closely with the three drivers of climate risk—hazards, exposure, and vulnerabilities—that the UN Intergovernmental Panel on Climate Change (IPCC) used as the basis for its climate risk analyses ten years later in preparing the two most recent assessment reports.[21] The same classification, subsequently updated with IPCC terminology, was applied in our later work throughout the 2020s.[22] I discuss this process in more detail—from the initial attempts around 2005 up to 2025—in the chapter on climate adaptation.

In the *transport sector*, I participated in the development of sustainability indicators targeting both passenger and freight transport. The first project was carried out for Oslo Sporveier, the largest public transport company in Norway. At the end of the 1990s, the company decided to shift its profile from "environmentally friendly" to "sustainable" mobility. They chose to ask WNRI for assistance rather than any of the many Oslo-based consulting firms or research environments (such as the Institute of Transport Economics) because of our critically oriented sustainability research. We conducted three rounds of developing an indicator-based sustainability accounting system for Oslo Sporveier (in 2000, 2003, and 2005).[23] The experiences from this work were later applied in a project for a working group under the Nordic Council of Ministers, developing an indicator system to assess the sustainability of freight transport—a system that was never implemented.[24]

---

[17] Farsund et al. (2001).

[18] Western Norway Research Institute (WNRI) (n.d.).

[19] Aall et al. (2012).

[20] Groven et al. (2006).

[21] Reisinger et al. (2020).

[22] Rød et al. (2025b).

[23] Sataøen and Andersen (2006).

[24] Aall and Langeland (2005).

In the *tourism sector*, development took a different course. Several of my colleagues and I were involved in establishing what would become an official national indicator system for sustainable tourism, but many of our suggestions for indicators were rejected. This included our proposal for an indicator addressing greenhouse gas emissions from transportation to and from Norway. We also criticized the design of indicators intended to address social sustainability for failing to capture the issue of fair distribution across time and space. However, we achieved significant influence with our input in developing a regional policy for sustainable tourism in our own home county of Sogn og Fjordane. I discuss this debate further in the chapter on sustainable tourism.

The Corruption of the Political Understanding of the Sustainability Goal

The second Norwegian White Paper on sustainable development, White Paper 58 (1996–1997), *Environmental Policy for Sustainable Development*, was presented by the then Minister of the Environment Torbjørn Berntsen. It provides a more mature and considered review of the sustainability goal and the policies that can be derived from it than the first report, which was published just two years after the World Commission's report. Although several white papers have been published since then, in my assessment, White Paper 58 offers the most comprehensive and revealing politically sanctioned discussion of the sustainability goal. The paper makes clear that the sustainability objective originates from two debates—the environmental debate and the developmental debate—and establishes three perspectives on the concept of sustainability:

- The ecological perspective: Preserve the biological diversity on Earth.
- The welfare perspective: Ensure that everyone today has equal rights to have their basic needs met.
- The generational perspective: Ensure that everyone in the future has equal rights to have their basic needs met.

The report also began with an important delimitation (p. 9, my emphasis)[25]:

> In the Long-Term Programme for 1998–2001, the Government affirms that sustainable development must be based on the welfare perspective, the ecological perspective, and the generational perspective. *The purpose of this white paper is to clarify and elaborate on the ecological perspective*, which must be viewed in close connection with the other two perspectives.

The understanding of the sustainability goal expressed in Parliamentary Report No. 58 (1996–1997) largely corresponds with the perspective developed at WNRI and ProSus, which may suggest that we had some influence on the political discourse

[25] Miljøverndepartementet (1997).

at that time. However, the political interpretation of the sustainability goal would eventually change significantly, both internationally and in Norway.

The introduction of the "triple bottom line" (TBL) marks the starting point of what I consider a problematic shift in the sustainability discourse. The management think tank AccountAbility coined the TBL-term in the mid-1990s, and John Elkington popularized it in his book *Cannibals with Forks—The Triple Bottom Line of twenty-first Century Business*. Elkington's commendable goal was to make the concept of sustainable development more accessible to the business community by presenting it as a third "bottom line," in addition to the long-established economic bottom line and the social bottom line. The latter originated from the corporate responsibility discourse of the 1960s and 1970s, addressing issues such as low wages and child labor.

The TBL approach to defining sustainability in business terminology became a success and eventually gained international acceptance through various guidelines, business charters, and so on. Over time, the TBL-term also attracted interest in public administration and policy development, and it was transferred back from the business sector to public governance and policymaking. Now, the three forms of "bottom line" are being replaced by three forms of sustainability, which we may refer to as "triple sustainability" (3-SD). However, there is an obvious risk associated with such transfer processes, as my colleague, Erling Holden, formulated (p. 29)[26]:

> There may be a significant difference between business objectives and the broader goals of society.

My colleague Karl Georg Høyer reinforces this point and argues that the TBL/3-SD term and the original Brundtland Commission's two-dimensional approach are fundamentally incompatible, not complementary.[27] The TBL frames sustainability as a balancing act between three distinct and separate spheres: economic, social, and ecological. In contrast, the Brundtland 2-SD concept emphasizes the integrative aspects of these dimensions.

On the journey from 2-SD, via TBL, and back to a 3-SD understanding of the sustainability goal, the interpretation of the goal changed so profoundly that we can speak of a perversion—a word meaning to turn upside down, to subvert, to distort. In public governance and policymaking, the term "economic bottom line" does not carry the same meaning as in private enterprise. The economic bottom line was therefore redefined (and renamed) as "economic sustainability," with its content linked to the traditional economic developmental concept of economic growth. Similarly, the social bottom line was renamed "social sustainability," moving away from the World Commission's original focus on the essential needs of the world's poor. In its place, the concept has been progressively expanded to encompass needs much higher up the hierarchy, effectively broadening the scope of "sustainability" to satisfy what are in reality affluent "wants" rather than vital needs.

---

[26] Holden (2007).

[27] Høyer (2010).

In my assessment, the direction of the dominant understanding of the sustainability goal can be characterized by borrowing a pointed formulation from the planning scholar Aaron Wildavsky. In his article, "If planning is everything, maybe it's nothing," Wildavsky argues that expectations for public planning had become filled with too many conflicting demands, rendering it unmanageable and causing it to lose its value as a tool for steering societal development.[28] My point, therefore, is that expectations for the goal of sustainable development have brought us into the same situation. The dominant understanding of the sustainability goal, as expressed in national policy and, to some extent, by the UN, means that the goal has, in reality, lost its value as a means of steering societal development in a different direction from the development that would have occurred without such a goal. As I will demonstrate in the following chapters, we must search for examples of an understanding of sustainability that aspires to guide development in a direction other than the very trajectory that led to the goal's formulation in the first place.

If the Sustainability Goal Is Everything—Then Perhaps It Is Nothing?

In 2020, I was invited by the Norwegian Association of Municipal Auditors to give a presentation on the sustainability goal at their annual conference for members of municipal audit committees. They wanted professional input in response to the growing demand for conducting performance audits in the area of sustainability. In my presentation, I asked, "If the sustainability goal is everything—then perhaps it is nothing?" What I did not know when preparing my talk was what the previous speaker, then Auditor General and former Conservative Party Minister Per-Kristian Foss, would say in his presentation. I was greatly surprised—or perhaps rather pleased—when I heard him dryly pose this ironic question: Are there, in fact, any areas of public policy that are not encompassed by the sustainability goals?[29]

A Shift from "Sustainable" to "Sustained" Development Indicators, and from "Political Governance" to "Administrative Reporting"

In 2005, three parties—the Labour Party, the Socialist Left Party (*Sosialistisk Venstreparti* [SV]), and the Centre Party (*Senterpartiet*)—secured a majority in the parliamentary election for the first time and formed a coalition government with Jens Stoltenberg (Labour Party) as prime minister. This government became known as the Stoltenberg II government. For the first time in history, SV joined a government.

---

[28] Wildavsky (1973).

[29] Norsk rikskringkasting (NRK) (2024c, September 25).

The then party leader of SV, Kristin Halvorsen, was additionally appointed to the second most powerful position in the government, serving as Minister of Finance. One of the issues SV promised to champion was strengthening sustainability efforts.

In the lead-up to Norway becoming the first country in the world in 1972 to establish a dedicated ministry with an associated minister responsible for environmental protection, a political struggle emerged. The question was whether this should become a superior ministry with the same authority over other ministries in environmental and natural resource matters as the Ministry of Finance holds over economic and state finance matters. The outcome, however, was a traditional sectoral ministry without significant governing authority over environmental issues relative to other sectoral ministries. In June 2006, the Ministry of Finance and Minister of Finance Kristin Halvorsen invited participants to a seminar that could be interpreted as a renewed debate on this issue. I was fortunate enough to attend.

The immediate background for the seminar was a proposal for a national set of sustainability indicators, presented the previous year by a public commission.[30] The previous government, known as the Stoltenberg I government—a minority government consisting of the Labour Party—had appointed the public commission. At the seminar, Minister of Finance Kristin Halvorsen announced that the Ministry of Finance would assume overall responsibility for sustainable development work from the Ministry of the Environment. This move received strong support from the environmental movement representatives present, but not from one of the political veterans of Norwegian environmental policy: former Minister of the Environment Torbjørn Berntsen from the Labour Party. He delivered a forceful political critique directed at the environmental movement participants, with reference to the economists in the Ministry of Finance[31]:

> I don't quite understand why you are so positive. Haven't you been paying attention? These are the very people who have been our main opponents in the environmental policy debates.

Berntsen was one of the few at the seminar who expressed skepticism toward Kristin Halvorsen's message. I shared this skepticism, based on two concerns. My first concern was the list of participants. No representatives from the political leadership of the Ministry of the Environment attended, nor did any members of the Environment and Energy Committee in the Storting. My second, and much greater, concern related to the content of the proposed set of indicators and the discussion surrounding some of them during the seminar.

One indicator intended to shed light on social sustainability measured the extent of imports to Norway from the world's least developed countries (the so-called LDC countries). At the seminar, we were presented with figures showing a positive trend over the past year. However, the representative from Statistics Norway had to admit that this growth was due to a large import of crude oil from one LDC country to the oil nation of Norway (!). At the seminar, then State Secretary in the Ministry of Foreign Affairs, Liv Monica Bargem Stubholt from the Centre Party, defended

---

[30] Finansdepartementet (2005).

[31] Personal notes taken during the meeting.

the indicator's relevance. She argued that increased consumption of coal, oil, and gas is a prerequisite for greater economic development, and thus also for increased sustainability. According to Stubholt, such development can be reconciled with the sustainability goal through a major commitment to carbon capture and storage. When I asked whether the government had a plan B if a carbon capture society proved impossible to achieve in practice, she dismissed the issue as irrelevant and replied, "We must believe in our goals." Two months later, Stubholt changed political posts and became State Secretary in the Ministry of Petroleum and Energy.

The topic raised most frequently during the Ministry of Finance's seminar was the "aging population," specifically linked to two proposed indicators: "the need for fiscal tightening as a share of gross national product" and "the proportion of long-term unemployed and recipients of disability benefits in the population." This is, of course, an important issue. However, the critical question is whether the "need for fiscal tightening" in public finances—in what was, in 2007, the world's third richest country—is truly a central sustainability issue. This question becomes particularly pressing when the same set of indicators does not include any measures comparing our consumption level with that of poorer countries.

One year later—in the autumn of 2007—the Ministry of Finance presented its final proposal for sustainability indicators, consisting of 18 indicators (see Table 1.4). Over time, a practice was established of annually updating the sustainability indicators, published in connection with the presentation of the national budget for the coming year.

In 2013, the government changed from the red-green Stoltenberg II coalition to a blue-green coalition consisting of the Conservative Party, the Liberal Party, the Christian Democratic Party, and the right-wing populist Progress Party (*Fremskrittspartiet*). That same year, the then-leader of the Progress Party, Siv Jensen, became Minister of Finance. One of her first actions upon taking office was to discontinue the practice of presenting an update on sustainability indicators alongside the national budget.

The system of annual reporting to the Storting on sustainability indicators was reinstated in 2020, the same year the Progress Party left the blue-green coalition government due to disagreements over immigration and asylum policy. It was a conservative finance minister who reintroduced the system. In 2021, Statistics Norway established a digital portal featuring the national indicators, where anyone can access and monitor Norway's progress toward achieving the sustainability goals.[32]

The 2021 version of the indicator system is markedly different from the 2007 edition. What was once a relatively *clear* and *simple* tool with fewer than 20 indicators, clearly adapted to a *Norwegian* context and designed to support *political governance*, now appears as a direct copy of the UN's indicator system. We are now faced with a *complex* and *unwieldy* tool. The number of indicators has increased more than tenfold compared to 2007—to 234—and the system is essentially a Norwegian

---

[32] Statistics Norway (SSB) (n.d.).

Table 1.4 First official version of national sustainability indicators for Norway presented by the Stoltenberg II government[33]

Sustainability theme	Sustainability indicators
Global justice	1. Official Norwegian aid, level and percentage of gross national income.
	2. Trade with least developed countries and developing countries combined.
Ecological conditions	3. Norwegian greenhouse gas emissions related to the Kyoto targets.
	4. Emissions of long-range transboundary air pollutants and emission commitments under the Gothenburg protocol.
	5. Population trends of breeding birds in terrestrial ecosystems.
	6. Proportion of water bodies with good or very good ecological status, freshwater.
	7. Proportion of water bodies with good or very good ecological status, coastal waters by region.
	8. The condition status of protected buildings.
	9. Total energy use per unit of gross national product.
	10. Estimated spawning stock for Northeast Arctic cod and Norwegian spring-spawning herring compared with precautionary reference points.
	11. Irreversible loss of productive land.
	12. Potential exposure to hazardous substances for health and the environment.
Economic conditions	13. Net national income per capita by source.
	14. Trends in income distribution.
	15. Generational accounts: Fiscal consolidation needs in public finances as a percentage of GDP.
	16. Population by highest level of education.
	17. Long-term unemployed and recipients of disability benefits as a proportion of the population.
	18. Life expectancy at birth.

translation of the UN's indicators with few or no adjustments to Norwegian conditions. The purpose of the indicator system appears to have shifted from political governance within the Norwegian political landscape to administrative reporting, primarily to the UN Secretariat for monitoring countries' progress on sustainable development: the High-Level Political Forum on Sustainable Development (HLPF).

The changes from 2007 to 2021 also involved a *thematic* shift. In line with the dual structure of the sustainability goal, it is reasonable to distinguish between the orientation of sub-goals and indicators toward the sustainability component (ecological sustainability) and the developmental component (social sustainability, which later came to include what is now termed economic sustainability). The World Commission's report emphasizes that both components are equally important in the sense that one cannot be achieved without the other. It is difficult to measure the emphasis

[33] Finansdepartementet (2007).

placed on these two aspects in the indicator system, but if we attempt to do so—by counting the number of indicators that can be wholly or partially allocated to each component—we arrive at a relatively clear picture of a value-based reorientation:

- 2007 indicator system ($n = 18$ indicators): 36% on ecological and 64% on social sustainability.
- 2021 indicator system ($n = 234$ indicators): 25% on ecological and 75% on social sustainability.

As noted above, such numerical exercises can only serve as an *indication* of the emphasis placed on ecological versus social sustainability. Nevertheless, the impression remains clear: *social* sustainability is evidently being given increasingly *greater* weight. The tendency for a country like Norway to place such significant—and apparently growing—emphasis on reporting social sustainability appears particularly problematic when compared to other countries, which almost without exception score lower than Norway on social sustainability indicators. In the annual Human Development Index (HDI) ranking used by the United Nations Development Programme (UNDP), Norway consistently ranks among the top three countries, together with the other Nordic nations and European countries such as Switzerland, the Netherlands, Germany, and Belgium.[34] At the same time, Norway has among the highest greenhouse gas emissions per capita. In fact, it has the highest per capita emissions in the world if emissions from the combustion of its exported oil and gas are included (see Fig. 1.1).

In another study conducted by the Norwegian Broadcasting Corporation (NRK) in collaboration with ten European media outlets, a senior researcher at the Norwegian Institute for Nature Research, and the organization Arena for Journalism in Europe, Norway also ranked very poorly on another key issue related to ecological sustainability. Data from 2025 show that Norway had the highest annual loss of area of natural and arable land per capita (see Fig. 1.2).

In other words, Norwegian authorities appear to have succumbed to the temptation to emphasize those aspects of sustainability efforts where Norway is best positioned to succeed—due to our economic history—while placing the least emphasis on the more problematic aspects related to ecological sustainability. The sociologist William Bruce Cameron is known for a quote that aptly captures this situation[35]:

> Not everything that can be measured matters, and not everything that matters can be measured.

But in contrast to Cameron's observation, which concerns methodological limitations, the situation in Norway is characterized by a political limitation: the replacement of the final "can" with "will be." The two most recent Norwegian parliamentary reports on sustainable development—from 2021 and 2025—address precisely the question of whether "everything that counts" is actually "counted." The first of these reports discusses two fundamentally different ways of measuring sustainability

---

[34] United Nations Development Programme (UNDP) (n.d.).

[35] Cameron (1963).

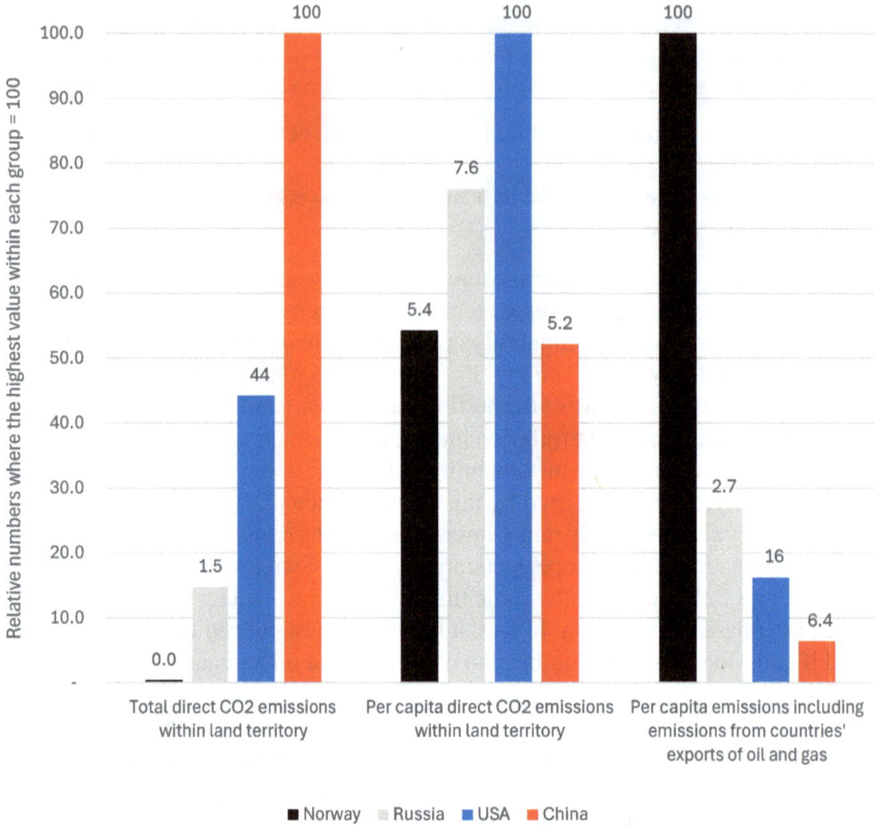

Fig. 1.1 National CO_2 emissions under three different system boundaries for Norway, Russia, the United States, and China in 2020 [36]

efforts: the SDG score and the spillover score. The latter reflects the extent to which a country's actions may have positive or negative effects on other countries' ability to achieve the SDGs, particularly through impacts on the environment and conditions via trade, economy and finance, and security. The report notes that Norway scores highly on the first (not least due to high scores on indicators related to social sustainability) and correspondingly low on the second. The report summarizes this as follows (p. 17)[37]:

> In the 2024 report, Norway is ranked with a score of 58.6 on a scale from 0 to 100, where 0 is the lowest. This is below the average for OECD countries, which stands at 73.8. Norway's low score can largely be attributed to negative effects associated with our high material consumption and extensive import of goods.

[36] Norsk rikskringkasting (NRK) (2021).

[37] Digitaliserings- og forvaltningsdepartementet (2025).

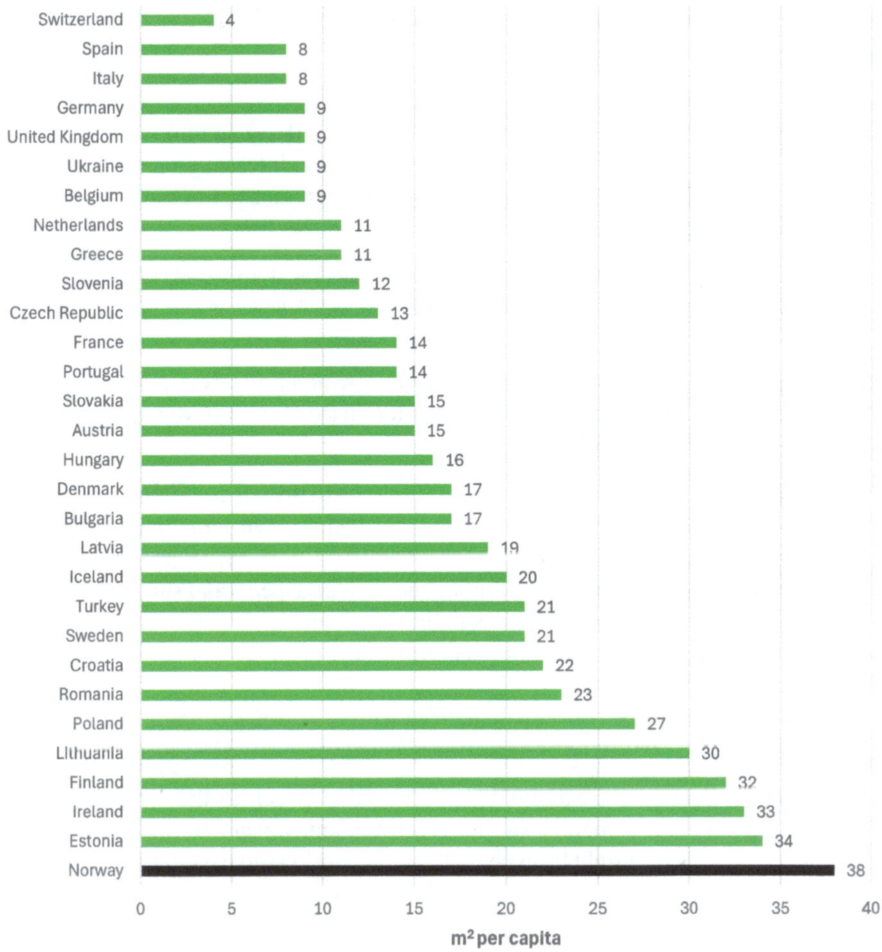

Fig. 1.2 Annual area of natural land and agricultural soil converted per capita[38]

In other words, we are reminded that the two fundamental sources of conflict inherent in the concept of sustainability—which several observers pointed out at the very beginning of the sustainability debate—still apply and will therefore continue to create problems[39]: An imprecise understanding of the relationship between global poverty and environmental degradation, and a conceptual ambiguity regarding the role of economic growth in relation to sustainable development.

---

[38] Norsk rikskringkasting (NRK) (2024a).

[39] Lele (1991).

Where Does the Sustainability Debate Stand Today, Internationally and in Norway?

My good friend, colleague, and mentor William M. Lafferty summarized his many years as a central participant in both the Norwegian and international sustainability debates in 2012. He compared the scientific, business, and political discourses on sustainable development (p. 13)[40]:

> The political discourse is much more negative: The UN-sponsored programme for promoting sustainable development has been seriously stymied at the international, regional, and national levels of implementation.

The developments from the 2000s to the present—which Lafferty considered particularly negative within the political sphere of the discourse—I have summarized in many lectures for my students in four points:

- Expanding from an original twofold concept consisting of ecological and social sustainability to a threefold concept encompassing environmental, social, and economic sustainability.
- Reduced willingness to prioritize among the various aspects and sub-goals related to the overarching objective of sustainable development.
- Insufficient communication of the need to accept absolute environmental limits on economic development.
- Broadening the scope of needs that must be addressed to achieve social sustainability.

Gro Harlem Brundtland was invited to deliver a speech to the United Nations General Assembly at the opening of the UN's 80th anniversary commemoration on September 22, 2025. As might be expected, she also addressed the goal of sustainable development, including the following reflection as reported by NRK[41]:

> But today we have exceeded the planet's boundaries, and we are approaching tipping points from which there is no return.

What Brundtland asserts can be substantiated by two comprehensive and internationally recognized studies concerning planetary boundaries and tipping point events.

Over several years, the Stockholm Resilience Centre has developed the *planetary boundaries* framework to highlight the increasing pressure on critical global processes that regulate the Earth's stability and resilience. In 2025, the research group has assessed that seven out of nine[42] critical Earth system boundaries have now been

[40] Lafferty (2012).

[41] Norsk rikskringkasting (NRK) (2024b).

[42] (1) climate change, (2) introduction of novel entities, (3) modification of biogeochemical flows, (4) freshwater change, (5) land system change, (6) change in biosphere integrity, (7) ocean acidification, (8) increase in atmospheric aerosol loading, and (9) stratospheric ozone depletion.

Table 1.5 Two prevailing interpretations of the sustainability goal (adapted from Aall, 2021) (see Footnote 12)

Characteristics	Broad and shallow understanding	Narrow and deep understanding
Sustainability theme and sub-goal	Many	Few
Prioritization between topics and sub-goals	Flat structure	Hierarchical structure
Quantification of sub-goals	Low weight	High weight
Decision criterion	Consensus and emphasis on win–win solutions	Majority decision and emphasis on trade-off situations
View on technology and social organization	Techno-optimistic (relying on a "technological fix"), production-focused, and uncritical of economic growth	Techno-skeptical (emphasizing systemic/societal change), consumption-focused, and critical of economic growth in wealthy countries

breached.[43] This is one more (ocean acidification) than the previous assessment (in 2023), leaving only two critical Earth systems still within the estimated safe zone (#8 increase in atmospheric aerosol loading, and #9 stratospheric ozone depletion).

Extensive studies have been conducted on so-called *global tipping point events*, which involve long-term and irreversible consequences of climate change that threaten humanity by disrupting life support systems and societal stability. The University of Exeter's Global Systems Institute leads the study, which includes over 200 researchers and practitioners from more than 25 institutions, including myself as a co-author of the 2025 edition of the chapter "Impact Governance for Earth System Tipping Points." The main conclusion from the second Global Tipping Point report is summarized as follows (p. 6)[44]:

> The world has entered a new reality. Global warming will soon exceed 1.5 °C. This places humanity in the danger zone, where multiple climate tipping points pose catastrophic risks to billions of people.

In a historical review I conducted in 2021 of the work on developing and implementing sustainability indicators in Norwegian municipalities and county authorities, I identified two main positions in the understanding of the sustainability goal: one broad and inclusive, and one narrow and focused (cf. Table 1.5). This categorization draws on similar dichotomies, such as "deep versus shallow," "strong versus weak," and "broad versus narrow" sustainability.

The Norwegian history of understanding the sustainability goal began with a narrow and specific interpretation in the 1990s, but has gradually shifted, in line with

---

[43] Sakschewski et al. (2025).

[44] Lenton et al. (2025).

international developments, toward a broader and more generalized understanding. The current challenge is to return to a narrower and more focused interpretation. Two indications suggest this may occur in the Norwegian debate, although I fear that the likelihood of this happening is less than the probability that we will continue along the broader and more generalized path:

- A growing interest in the concept of *circular economy.*
- An increasing number of contributions to the public debate *criticizing* the goal of *economic growth.*

The concept of circularity has always held a central place in natural resource management within Norwegian environmental policy. As part of the preparatory work for establishing the Ministry of the Environment in 1972, creating a national natural resource accounting system—parallel to the existing national economic accounts—was high on the agenda and was described in a dedicated government report.[45] The ambitions for such a system were eventually scaled back. Although today's Norwegian sustainability indicator system has historical roots in the original ideas surrounding a national natural resource accounting framework, its design has become markedly different, as described above. Nevertheless, the original intent behind the resource accounting system continues to leave a clear mark on Norwegian environmental policy through the mitigation hierarchy principle. This principle dictates that the first priority in natural resource management is to "reduce," followed by "substitute," and finally "improve efficiency." The principle has found its clearest expression in waste policy, where reduction of waste volume is prioritized over recycling, which in turn is prioritized over the final disposal of waste. Within recycling, reuse is prioritized over material recovery or energy recovery. Parallels to such a mitigation hierarchy can also be found in transport and environmental policy (reducing mobility is prioritized over shifting from private to public transport, which in turn is prioritized over improving the efficiency of individual modes of transport). Other policy areas within the environmental sector have also adopted a similar approach.

The modern approach to the circular economy, like much else in Norwegian environmental policy since 2000, originated from the EU, which adopted its First Circular Economy Action Plan in 2015. Norway incorporated the principle two years later in a white paper on waste policy. In 2021, the government followed up with the first national strategy for the circular economy, which was succeeded by a dedicated white paper three years later. This white paper introduced the requirement that the government regularly develop and update action plans for the circular economy. Although this appears to be a comprehensive commitment to the circular economy, it remains to be seen whether practical policy will steer Norway onto a different, narrower, and more focused path toward sustainability. The 2021 white paper, which set the overall political direction for further efforts, states that circularity is a political tool to maintain and increase economic growth, albeit with some pious hopes that "economic growth can occur without increased use of resources" (p. 6).[46]

---

[45] Miljøverndepartementet (1977).

[46] Klima- og miljødepartementet (2021).

I have greater confidence in the resurgence of the 1970s public criticism of economic growth. This resurgence is linked to, among other things, the idea of "Doughnut Economics," first formulated in 2012 by Kate Raworth in the Oxfam report *A Safe and Just Space for Humanity: Can we live within the doughnut?* and later developed further in her book *Doughnut Economics: Seven Ways to Think Like a twenty-first-Century Economist*, published five years later. Raworth uses the doughnut as a visual model: the inner ring illustrates the social foundation (basic needs that must be met), and the outer ring represents the planet's ecological ceiling (the environmental boundaries we must not exceed). Environmental organizations and professional networks frequently use the model in Norwegian public debate to advance this way of thinking, but it has thus far not gained traction in party-political discourse. The similarity to Karl Georg Høyer's hierarchical understanding of sustainability (see Table 1.1), developed two decades before the Doughnut model, demonstrates—if nothing else—that the research group at Western Norway Research Institute was on a sensible track at the time. It remains to be seen whether Norwegian sustainability policy can return to that same path. In the following chapters, I will review how various policy areas I have researched over the years have alternated between the two sustainability approaches—broad/shallow and narrow/deep—as summarized in Table 1.5.

Norwegian Participation at a Mere 1% at the Fourth International Degrowth Conference in 2014

In 2008, the first international conference on "Economic De-Growth for Ecological Sustainability and Social Equity" was held in Paris. This conference and subsequent degrowth conferences were based on degrowth ideas formulated by, among others, the economists Nicholas Georgescu-Roegen (a key reference in my master's thesis at the Norwegian University of Life Sciences) and Joachim Spangenberg (whose work we relied on when developing sustainability indicators). However, it would be six years before I finally made it to one of these conferences. I attended the fourth in the series, held in Leipzig, together with my colleague Hans Jakob Walnum. We were the only two Norwegians among the 2709 registered delegates.

But we were more than just delegates. We had been accepted to host a parallel seminar on rebound mechanisms, conducted in collaboration between Western Norway Research Institute (the two of us) and Tilman Santarius at the Technical University of Berlin. The invited presentations from the seminar were later included in the book *Rethinking Climate and Energy Policies: New Perspectives on the Rebound Phenomenon*, edited by Hans Jakob, Tilman, and me, with contributions from 17 other researchers.

References

Aall, C. (2014). Sustainable tourism in practice: Promoting or perverting the quest for a sustainable development? *Sustainability, 6*, 2562–2583. https://doi.org/10.3390/su6052562

Aall, C. (2021). *Bærekraftindikatorer. Et blikk bakover—og framover—ment å være til nytte og inspirasjon for kommuner og fylkeskommuner i arbeidet med å realisere FNs bærekraft*mål [Sustainability indicators. A look back—and forward—for the realization of the UN Sustainable Development Goals] (VF-report 7/2021). Western Norway Research Institute.

Aall, C., & Langeland, A. (2005). *Nordisk Transportpolitisk Nettverk (NTN): Indikatorer for bærekraftig godstransport* [Nordic Transport Policy Network: Indicators for sustainable freight transport] (VF-report 3/05). Western Norway Research Institute/University of Stavanger.

Aall, C., & Norland, I. (2002). *Det økologiske fotavtrykk for Oslo kommune - resultater og forslag til anvendelse av økologisk fotavtrykk som styringsindikat*or [The ecological footprint for Oslo municipality] (Report 2/2002). Western Norway Research Institute/ProSus.

Aall, C., & Norland, I. (2005). The use of the 'ecological footprint' in local politics and administration: Results and implications from Norway. *Local Environment, 10*(2), 159–172. https://doi.org/10.1080/13549830500042453

Aall, C., & Norland, I. T. (2003). *Indikatorer for vurdering av lokal klimasårbarhet* [Indicators for assessing local climate vulnerability] (VF-report 15/03). Western Norway Research Institute.

Aall, C., Breisnes, K., Hille, J., & Høyer, K. G. (2002). *Bærekraftige kommuner i praksis. Omtale av et styringssystem for integrering og konkretisering av bærekraftig praksis i kommunen* [Sustainable municipalities in practice] (VF-report 17/02). Western Norway Research Institute.

Aall, C., Heiberg, E., Simonsen, M., Walnum, H. J., & Ekström, F. (2012). *Korleis vurdere effekten av ein regional klimapolitikk?* [How to assess the effect of a regional climate policy?] (VF-report 7/2012). Western Norway Research Institute.

Cameron, W. B. (1963). *Informal sociology: A casual introduction to sociological thinking*. Random House.

Digitaliserings- og forvaltningsdepartementet. (2025). *Norges arbeid med bærekraftmålene. Status, utfordringer og veien videre* [Norway's work with the Sustainable Development Goals] (Meld. St. 35 [2024–2025]).

Farsund, A., Hille, J., & Aall, C. (2001). *Klima- og energiplan for Stavanger kommune* [Climate and energy plan for Stavanger municipality] (Report RF 2001/184). Rogalandsforskning.

Finansdepartementet. (2005). *Enkle signaler i en kompleks verden. Forslag til et nasjonalt indikatorsett for bærekraftig utvikling* [Simple signals in a complex world] (NOU 2005:5).

Finansdepartementet. (2007). *Norges strategi for bærekraftig utvikling* [Norway's strategy for sustainable development].

Groven, K., Sataøen, H. L., & Aall, C. (2006). *Regional klimasårbarheitsanalyse for Nord-Norge* [Regional climate vulnerability analysis for Northern Norway] (VF-report 4/06). Western Norway Research Institute.

Heiberg, E., Aall, C., Amundsen, H., Storm, H., Høyer, K. G., Næss, L. O., Solstad, S. P., & Hovelsrud, G. K. (2008). *Indikatorer for lokale klimasårbarhetsanalyser. Kunnskapsstatus og skisse til en metode for utprøving i norske kommuner* [Indicators for local climate vulnerability analyses] (VF-report 5/08). Western Norway Research Institute.

Holden, E. (2007). *Achieving sustainable mobility: Everyday and leisure-time travel in the EU*. Ashgate Publishing.

Høyer, K. G. (1991). Regionalpolitikkens økopolitiske grunnlag [The eco-political basis of regional policy]. In K. G. Høyer & T. Selstad (Eds.), *Regionalpolitikkens økologiske grunnlag* (pp. 5–48). Western Norway Research Institute/NordREFO.

Høyer, K. G. (1997). Sustainable development. In D. Brune, D. Chapman, & M. Gwynne (Eds.), *The global environment* (pp. 1185–1208). VCH Publ.

Høyer, K. G. (2010). Seven theses on CO_2-reductionism and its interdisciplinary counteraction. In R. Bhaskar, C. Frank, K. G. Høyer, P. Næss, & J. Parker (Eds.), *Interdisciplinarity and climate change* (pp. 35–54). Routledge.

Høyer, K. G., & Aall, C. (1997). *Miljø- og bærekraftindikatorer. En internasjonal kunnskapsoversikt som grunnlag for utvikling av en retningsanalysemodell* [Environmental and sustainability indicators. An international knowledge review] (Report 13/97). Western Norway Research Institute.

Klima- og miljødepartementet. (2021). *Nasjonal strategi for ein grøn, sirkulær økonomi* [National strategy for a green, circular economy].

Lafferty, W. M. (2012). *Time and sustainable development* [Paper presentation]. TARDIS 2012, Leibnitz, Austria. http://tardis.tugraz.at/wiki/lib/exe/fetch.php?media=tardis2012_01_lafferty.pdf

Lafferty, W. M., & Langhelle, O. (Eds.). (1999). *Towards sustainable development: The goals of development—and the conditions of sustainability.* Macmillan/St. Martin's Press.

Lele, S. M. (1991). Sustainable development: A critical review. *World Development, 19*(6), 607–621. https://doi.org/10.1016/0305-750X(91)90197-P

Lenton, T. M., Milkoreit, M., Willcock, S., Abrams, J. F., Armstrong McKay, D. I., Buxton, J. E., Donges, J. F., Loriani, S., Wunderling, N., Alkemade, F., Barrett, M., Constantino, S., Powell, T., Smith, S. R., Boulton, C. A., Dennis, D., Dijkstra, H., Pearce-Kelly, P., Pinho, P., & Roman-Cuesta, R. M. (Eds.). (2025). *The Global Tipping Points Report 2025.* University of Exeter.

MacGillivray, A., & Zadek, S. (Eds.). (1995). *Accounting for change. Papers from an international seminar.* The New Economics Foundation.

Meadows, D. H. (1993, May 20). Seattle citizens define their own Dow-Jones average. *The Global Citizen.* https://donellameadows.org/archives/seattle-citizens-define-their-own-dow-jones-average/

Miljøverndepartementet. (1977). *Om arbeidet med et informasjonssystem for naturressurser* [Natural Resources Accounting] (NOU 1977:31).

Miljøverndepartementet. (1989). *Miljø og utvikling. Norges oppfølging av Verdenskommisjonens rapport* [Environment and development. Norway's follow-up of the World Commission's report] (Meld. St. nr. 46 [1988–89]).

Miljøverndepartementet. (1997). *Miljøvernpolitikk for en bærekraftig utvikling dugnad for framtida* [Environmental Policy for Sustainable Development—A joint effort for the future] (Meld. St. nr. 58 [1996–97]).

Næss, L. O., Norland, I., Lafferty, W., & Aall, C. (2006). Data and processes linking vulnerability assessment to adaptation decision-making on climate change in Norway. *Global Environmental Change, 16*(2), 221–233. https://doi.org/10.1016/j.gloenvcha.2006.01.007

Norsk rikskringkasting (NRK). (2021). *Norske CO_2-utslipp: Slik går Norge fra å være miniputt til gigant på klimastatistikken.* Retrieved March 8, 2026, from https://www.nrk.no/klima/xl/norske-co_-utslipp_-slik-gar-norge-fra-a-vaere-miniputt-til-gigant-pa-klimastatistikken-1.15600846

Norsk rikskringkasting (NRK). (2024a). *Europa fra grønt til grått: Nordmenn bygger ned mest natur.* Retrieved March 8, 2026, from https://www.nrk.no/dokumentar/xl/europa-fra-gront-til-gratt_-nordmenn-bygger-ned-mest-natur_-viser-europeisk-kartlegging-1.17577504

Norsk rikskringkasting (NRK). (2024b). *Gro Harlem Brundtland talte på FNs 80-årsjubileum.* Retrieved March 8, 2026, from https://www.nrk.no/urix/gro-harlem-brundtland-talte-pa-fns-80-arsjubileum-1.17580829

Norsk rikskringkasting (NRK). (2024c, September 25). *Guro Harlem Brundtland talte på FNs 80-årsjubileum* [Video]. YouTube. https://www.youtube.com/watch?v=GgkzRZWcTDM

Reisinger, A., Howden, M., Vera, C., Garschagen, M., Mach, K. J., Pathak, M., Poloczanska, E., van Aalst, M., Ruane, A. C., Hurlbert, M., Mintenbeck, K., Pedace, R., Rojas Corradi, M., Viner, D., Kreibiehl, S., O'Neill, B., Pörtner, H.-O., Sillmann, J., Jones, R., & Ranasinghe, R. (2020). *The concept of risk in the IPCC Sixth Assessment Report: A Summary of Cross-Working Group Discussions.* IPCC.

Rød, J. K., Aall, C., & Selseng, T. (2025a). Towards a holistic climate service providing information about four climate risk determinants. *Climate Services, 38,* 100,558. https://doi.org/10.1016/j.cliser.2024.100558

Rød, J. K., Aall, C., & Selseng, T. (2025b). Towards a holistic climate service providing information about four climate risk determinants. *Climate Services, 38*, 100,558. https://doi.org/10.1016/j.cliser.2024.100558

Sakschewski, B., Caesar, L., Andersen, L., Bechthold, M., Bergfeld, L., Beusen, A., Billing, M., Bodirsky, B. L., Botsyun, S., Dennis, D., Donges, J. F., Dou, X., Eriksson, A., Fetzer, I., Gerten, D., Häyhä, T., Hebden, S., Heckmann, T., Heilemann, A., et al. (2025). *Planetary Health Check 2025*. Potsdam Institute for Climate Impact Research.

Sataøen, H. L., & Andersen, O. (2006). *Samfunnsregnskap for Oslo Sporveier 2005* [Social accounting for Oslo Sporveier 2005] (VF-report 8/06). Western Norway Research Institute.

Statistics Norway (SSB). (n.d.). *Bærekraftsmålene* [The Sustainable Development Goals]. Retrieved March 8, 2026, from https://www.ssb.no/baerekraftsmaalene

United Nations Development Programme (UNDP). (n.d.). *Human Development Index*. Retrieved March 8, 2026, from https://hdr.undp.org/data-center/human-development-index#/indicies/HDI

Walnum, H. J., Aall, C., & Løkke, S. (2014). Can rebound effects explain why sustainable mobility has not been achieved? *Sustainability, 6*(12), 9510–9537. https://doi.org/10.3390/su6129510

Western Norway Research Institute (WNRI). (n.d.). *Miljøkalk*. https://vfp1.vestforsk.no/miljokalk/

Wildavsky, A. (1973). If planning is everything, maybe it's nothing. *Policy Sciences, 4*, 127–153. https://doi.org/10.1007/BF01405734

World Commission on Environment and Development (WCED). (1987). *Our common future*. Oxford University Press.

Xue, J., Walnum, H. J., Aall, C., & Næss, P. (2017). Two contrasting scenarios for a zero-emission future in a high-consumption society. *Sustainability, 9*(1), 20. https://doi.org/10.3390/su9010020

Chapter 2
Municipal Environmental Policy—From Government-Initiated Reform to Local Counter-Power

Abstract This chapter analyzes the emergence and development of municipal environmental policy in Norway from the 1980s to the present. The author takes as a starting point the transition from the pre-institutional phase to major state reforms, such as the Environmental Protection in Municipalities (*Miljøvern i kommunen* [MIK]) program, and explores the tension between state control and local autonomy. Drawing on personal experience as both an environmental manager and a researcher in major projects on municipal environmental work, the chapter highlights how Norwegian municipalities have acted both as implementers and as important countervoices in the national sustainability debate. The chapter also discusses the modern "dilution" of the concept of sustainability, where broad social and economic objectives increasingly compete with the original environmental ambitions, and questions what is required for the municipal sector to once again become a driving force in the pursuit of sustainable development.

Introduction

The massive bloom of the planktonic alga *Chrysochromulina polylepis* along the coast of southern Norway in 1988 strengthened the political basis for a demand long advocated for by both environmental organizations and government pollution authorities. They had called for increased municipal efforts to reduce local nutrient discharges, particularly a large-scale expansion of municipal wastewater treatment. This event also helped the Ministry of the Environment—despite its previous reputation as more of an "opponent" than a "partner" to the municipal sector—establish a productive collaboration with the Norwegian Association of Local and Regional Authorities (KS) from 1988 on the MIK program. An unintended consequence of the MIK initiative was the revelation of a "political cleavage" between the state and municipalities in achieving sustainable development.

The term "political cleavage" has traditionally described historically rooted and entrenched divisions between groups of citizens associated with economically, socially, or culturally defined classes, which in turn manifest as divisions

C. Aall, *Sustainable Development*, https://doi.org/10.1007/978-3-032-35930-8_2

between political parties.[1] Over time, new meanings of the term have emerged, often employing the less sharply defined concept of "political divide."[2] In the context of sustainability and climate, the term has also characterized divisions between levels of government within *multilevel governance* systems. Researchers use it to show how pioneering municipalities are not only "ahead" of other municipalities, but also "ahead" of their respective national authorities.[3] This also encompasses the phenomenon of these same pioneering municipalities collaborating internationally in climate negotiations through organizations such as the International Council of Local Environmental Initiatives (ICLEI) to advocate for more ambitious climate commitments than those supported by their national governments.[4]

Most liberal democracies have established environmental policy structures at the subnational level. What is distinctive about the Norwegian case is the national authorities' substantial commitment to initiating a local environmental policy reform. The state initially provided full funding for an environmental manager position in a three-year program covering 91 out of 435 municipalities at the time. Subsequently, from 1991 to 1997, a scheme was introduced whereby all municipalities with more than 2000 inhabitants were offered full funding for such a position. Municipalities with fewer than 2000 inhabitants were offered funding for a 50% position. By 1996, 420 out of 453 municipalities had accepted the scheme with earmarked funding for an environmental manager position.

In this chapter, I review the history from experimentation to reform and on to the consolidation phase, demonstrating how the municipal sector in general, and pioneering municipalities in particular, became a countervoice in the Norwegian sustainability debate. To illuminate this history, I draw especially on experiences from my first research project—the Eco-Municipality Programme, which I was appointed to lead—as well as later research on Local Agenda 21 and close collaboration with Professor William M. Lafferty and the Programme for Research and Documentation for a Sustainable Society (ProSus).

In conclusion, I discuss the current situation of Norwegian municipalities, many of which have eliminated the post of environmental officer. The concept of sustainability has evolved to encompass nearly all forms of policy, and local environmental policy has largely been reduced to local climate policy. I also discuss what is required for the municipal sector to once again contribute to sustainable development.

---

[1] Lipset and Rokkan (1967).

[2] Deegan-Krause (2007).

[3] Lafferty and Eckerberg (1998).

[4] Lindseth (2004).

An Environmental Policy Backdrop: The Pre-institutional Period

In an article in the first book on municipal environmental policy in Norwegian municipalities, I distinguish between four phases in the development of municipal environmental policy structures[5]:

- The *pre-institutional* period (up to 1985).
- The *experimental* period (1985–1992).
- The *reform* period (1992–1997).
- The *consolidation* period (after 1997).

Until the experimental period began in the mid-1980s, municipal environmental policy primarily addressed typical local issues related to water supply and sewage—two areas for which municipalities had already been assigned responsibility in 1860 through the Public Health Act. A hundred years later, with the Building Act of 1965, municipalities were also given a central role in land-use planning. Up to this point, the situation for Norwegian municipalities did not differ from that in many other European countries. However, the Norwegian Building Act introduced a unique element that would eventually distinguish Norway from other countries and prove significant for the subsequent phase of institutionalizing local environmental policy.

Simultaneously with the introduction of the Building Act, national political responsibility for nature and environmental policy was transferred from the Ministry of Church, Education, and Research to the Ministry of Local Government and Labour, which was responsible for regional and rural development and the implementation of regional and municipal planning. Seven years later, during the process of establishing the Ministry of the Environment, the connection between nature and environmental policy and planning once again became a topic of debate. Ultimately, responsibility for what in 1972 was called the Planning and Building Act was assigned to the newly established Ministry of the Environment.

The connection between nature and environmental policy and local planning under the Planning and Building Act emerged during a period when national environmental policy was being institutionalized in most Western democracies. This development was distinctive for Norway and diverged from the traditional view of nature and environmental policy as an exclusively national concern.[6] Most countries at the time focused on protecting nature from human intervention, whereas the Norwegian model emphasized that effective environmental management arises from a balance between use and conservation, with the political responsibility at the national level for ensuring this balance assigned to the ministry responsible for environmental policy. This perspective was intertwined with what eventually developed into a uniquely Norwegian form of green populism and eco-philosophy.[7] Consequently, the stages leading

---

[5] Aall (2002).

[6] Høyer and Selstad (1991).

[7] Kvaløy (1979).

up to the institutionalization of local environmental policy combined conventional land-use planning with a more radical, deep ecology-inspired approach.[8]

Initial Attempts to Strengthen Structures for Local Environmental Policy

During the 1970s, following the establishment of the Ministry of the Environment, a perception quickly developed within the ministry that many municipalities lacked sufficient expertise to manage their assigned environmental protection tasks adequately. The ministry also found it problematic that many environmental responsibilities fell into a gray area between the jurisdictions of various municipal agencies.[9] As early as 1980, calls emerged to involve municipalities and county authorities more actively in environmental protection efforts—specifically in Parliamentary Report 68 (1980–1981), *Protection of Norwegian Nature* and Official Norwegian Report 1980:23, *Nature Conservation in Norway*. In 1985, the first initiatives began with state grants for municipal environmental consultants, eventually covering twelve municipalities. It is the environmental issues related to nature conservation, the management of the commons, and agriculture that dominate these early attempts, rather than the conventional pollution control tasks that the Ministry of Environment had primarily concerned itself with until then (see Footnote 5).

During the 1980s, examples began to emerge—parallel to and independent of these initiatives—of municipalities engaging in areas that extended even further beyond the conventional boundaries of municipal environmental policy. One clear example, which eventually gained significant support, was the involvement of Norwegian municipalities and county authorities in opposing plans for a plutonium reprocessing plant in Dounreay, northern Scotland.[10] The Norwegian Ministry of Foreign Affairs even challenged this campaign on the grounds that municipalities should not interfere in foreign policy matters. However, the municipalities did not heed the ministry's objections. Their local activism eventually pressured national authorities to engage with the British government, which in turn contributed to the latter's decision in 1989 to halt further planning of the reprocessing facility. This intervention also brought the issue of radioactive pollution onto the local agenda, helping to lay the foundation for widespread local support for measures to address the extensive radioactive fallout in Norway from the 1986 Chernobyl accident, the world's most severe nuclear power accident to date (see Footnote 8).

In 1987, former Minister of the Environment and then Prime Minister Gro Harlem Brundtland (Labour Party) presented the UN report *Our Common Future*. As mentioned in the previous chapter, the Ministry of the Environment sent the report

[8] Aall (2000).

[9] Nøttestad (1999).

[10] Aall (1988).

for public consultation. Of the 167 responses received, 43 came from municipalities, four of which would later join the Norwegian Eco-Municipality Programme (Lardal, Nore og Uvdal, Sogndal, and Ølen), see below. The ministry's summary of the municipal feedback states, among other things (p. 171)[11]:

> The report has been received very positively overall, as a necessary wake-up call that must have implications for how we organize ourselves at the local level as well.

The municipal statements reflect an attitude toward local measures and changes that differs markedly from the prevailing perception at the time of what municipalities were concerned with regarding nature and environmental policy. The municipalities highlighted the following points, among others:

- The importance of the precautionary principle.
- Prioritization of energy conservation with the aim of reducing energy consumption in accordance with the recommendations of the Brundtland Report.
- Improved local control and monitoring.
- The "polluter pays" principle must apply across all sectors.
- Environmental crime should be followed up and penalized more severely.
- Clarification that a decision in favor of conservation issues represents a commitment to more vital values and interests.
- Incentivizing a reduction in industrial agriculture.
- Continued enforcement of strict farmland protection.
- Use of ecological accounting systems and budgets with sustainable development as a guiding premise.

All respondents emphasized the importance of strengthening environmental efforts at the local level, and several municipalities pointed out that the report insufficiently addressed the relationship between growth and conservation. Nevertheless, in Parliamentary Report 46 (1988–1989) on the follow-up on recommendations from the World Commission's report, the role of municipalities in environmental policy was primarily described as implementing national environmental policy to address distinctly local environmental issues. The report also noted, however, that municipalities must assume their share of responsibility for fulfilling Norway's international emission reduction commitments (p. 145) (see Footnote 11).

Preparations for the Major Reform Initiative Begin

1988 marked the year the major reform of municipal environmental policy began, abbreviated as MIK in Norwegian (Miljøvern i kommunene). It started as a pilot program involving 91 municipalities (until 1991) and then transitioned into a reform that included measures such as offering earmarked grants to finance a Municipal

---

[11] Miljøverndepartementet (1988).

Environmental Officer position in all Norwegian municipalities from 1992 until the scheme was fased out in 1997.

As a recent graduate in natural resource management from the Agricultural University of Norway, I had the pleasure of participating in this journey from the very beginning. I worked for two years as an employee in a municipality involved in the MIK program (Ølen municipality), and from 1990 as a researcher at Western Norway Research Institute (WNRI), responsible for developing research on municipal environmental policy.

The MIK program generated great enthusiasm—both from state authorities (or more precisely, from certain individuals within the Ministry of the Environment) and from the municipalities (and here as well, from specific individuals, particularly the 91 environmental managers). The Ministry of the Environment set a condition for participation: the environmental managers should receive high salaries (something I, as a recent graduate, benefited from), preferably at the level of department heads. The ministry also recommended (but did not require) that the environmental manager be included in the municipality's administrative leadership group. (I was not granted access to that group in my municipality.) The idea was that these two measures would ensure greater influence for environmental issues within the municipal organization.

When the program was launched, the 91 positions were advertised more or less simultaneously. The job announcement required higher education, natural science expertise—preferably combined with social science competence—and experience in municipal planning and administration. Candidates with this combination proved extremely difficult to recruit. Few university-level programs offered such a focus at the time—in reality, only one: the Nature Management program at the Agricultural University of Norway, established in 1972 after the creation of the Ministry of the Environment. Consequently, a large proportion of the newly hired were recent graduates in nature protection management. They had highly relevant education but little or no experience in municipal administration, and therefore often had correspondingly limited influence in internal administrative processes. I reflected all these characteristics.

"Don't Think You're Somebody!"

In 1988, I joined the ranks of individuals with relevant academic qualifications but no experience in municipal administration. I quickly felt the consequences. I struggled to engage with the prevailing administrative processes. During the two years I worked in the municipality, I was involved in land-use planning only once, and that was to provide a supporting statement in a case concerning the demolition of an illegally constructed veranda on a cabin in the coastal zone. The technical director was reluctant to physically enforce the decision and clearly needed all the support he could get.

A year after I started in the position, I was contacted by the Department of Geography at the University of Bergen. They were organizing a Regional Postgraduate Course in Spatial and Environmental Planning ("RegioPlan Vest").

The course coordinator invited me, explaining that they were particularly eager to include one of the newly appointed municipal environmental officers that year. I accepted, motivated by a desire to strengthen my competence in municipal planning. I applied to the municipality for paid study leave. I assumed this would be unproblematic, since the municipality would be reimbursed for salary expenses by the Ministry of the Environment. My application for leave was approved, but I was denied paid leave. The municipal administration committee, which processed my application, added the following to their formal decision: "Furthermore, we thought we had hired a highly competent person."

The MIK program generated significant networking activity. The Ministry of the Environment initiated county-level MIK forums, which held regular meetings between participating municipalities, the county administration, and regional state authorities—in addition to annual national gatherings. We also gained access to an early digital version of email and chat: a Swedish platform-independent communication system called PortaCom. This was likely the world's first nation wide digital environmental protection network. It enabled all MIK employees with computer access to communicate with each other and with a network of experts at various universities and research institutes. We could pose questions to "everyone" and share experiences about how to address specific environmental challenges. We also established our own professional interest organization (Forum for Local Environmental Policy, abbreviated as FLM in Norwegian) and created an internal newsletter, which was later taken over by the Ministry of the Environment and given a more professional design (*MIK-nytt*). My close colleague and friend in the neighboring municipality of Karmøy, Ole Jørgen Grann, played a key role in both initiatives. He served as secretary general for the FLM and as editor of *MIK-nytt*.

On How the Forum for Local Environmental Policy Was Almost Taken Over by the Communists!

I was among five initiators who founded an association for municipal environmental officers. The organization eventually received the rather prosaic name "Forum for Local Environmental Policy" (FLM). For some, that might have evoked revolutionary associations with liberation movements of the 1970s.

While planning the FLM's establishment, we received a small grant from the Ministry of the Environment to cover travel expenses for meeting in Oslo to discuss the organization's purpose and structure. That's where the problems began. Two of the five initiators had backgrounds in AKP-ml, at the time still an active and almost elitist communist party with the full name "The Workers' Communist Party (Marxist–Leninist)." In the 1970s, AKP-ml was known and feared for its unreserved support of Marxist liberation groups with acronyms reminiscent of our FLM, as well as for its repeated attempts to take over various organizations (typically trade unions, peace organizations, and solidarity groups) to use in advancing their party strategies of revolutionary communism inspired by Mao Zedong, Joseph Stalin, and classical Marxism–Leninism.

We experienced an echo of this strategy when the two former AKP sympathizers tried to turn the FLM into a revolutionary organization for environmental protection. They also wanted the FLM to become a trade union. We three other initiators were novices in this kind of organizational struggle. One of us—a calm former agricultural bureaucrat—even began to cry during the meeting when the discussion was at its most heated. But democracy prevailed. We were three against two and adopted a completely harmless, non-revolutionary organizational model. Few—if any—of the other 86 municipal environmental officers ever learned of this episode.

A few years later, after I had started at WNRI, I received an assignment from one of the two former AKP members. He was still working in one of the original MIK municipalities. Otherwise, he seemed a very pleasant and reasonable man. A week before I was to deliver the report, I saw in the newspaper that he had been summarily dismissed for criticizing a major industrial company for its lack of environmental efforts. However, he was reinstated for two hours to receive my report.

At times, it's a tough calling to be both an environmental bureaucrat and an environmental researcher, even outside the big cities!

The MIK program underwent a comprehensive evaluation involving nine research institutions, including WNRI. The KS feared the reform would give the state a new, extended reach into the municipalities. The KS also worried that the reform would increase municipal resource use in areas it originally considered to be state responsibilities. The evaluation showed these concerns were unfounded. Participation in the MIK program did not change the use of measures and instruments for which municipalities were responsible by law or tradition. On the contrary, the MIK program led to a proliferation of new and "soft" areas of intervention in the form of various information and awareness-raising initiatives. The program also resulted in a significant increase in environmental planning activities. Municipalities produced environmental status reports, formulated environmental objectives and developed action plans: the so-called Environmental and Natural Resource Programmes.

The evaluation produced two central findings regarding substantive policy content. First, over time, municipal environmental issues could not diverge significantly from the environmental concerns prevailing in the local community. Second, participation in the MIK program contributed to efforts to develop a municipal environmental management approach rooted in local objectives for managing natural resources within the municipalities. This offered an alternative to an approach based solely on national government standards.[12] Experiences from the MIK program demonstrated that municipalities wish to develop their own profile in environmental management; they do not want to act solely as implementers of national environmental policy.[13]

The Right Person in the Right Place

An important part of the MIK program was the regular national gatherings held across the country for all 91 municipal environmental officers, with participation from research communities and national environmental authorities. Another key aspect was a follow-up evaluation led by one of the leading figures in Norwegian social science research: Professor Alf-Inge Jansen at the Department of Administration and Organization Theory, University of Bergen. At one of the first national gatherings, in the autumn of 1988, Alf-Inge Jansen presented the evaluation framework. He began with a paraphrase of one of Winston Churchill's most famous quotes: "Never have so many people been placed in the right place at the right time." Jansen explained this by emphasizing the significance of the MIK reform and the role we, as environmental professionals, could play in shaping Norwegian environmental policy. He spoke with considerable authority and credibility, as that same year he submitted his doctoral dissertation titled "Structure and Agency: A Study of the Formulation of State Nature and Environmental Policy."

The First Two Research Projects Aimed at Examining Municipalities' Room for Manoeuvre in Their Efforts Toward Sustainable Development

WNRI played a central role in the first two research projects that explored the scope of action available to municipalities in their efforts toward sustainable development: the project Nature and Environmentally Friendly Urban Development (NAMIT) and the Eco-Municipality Programme. I was given responsibility for leading the latter as a newly appointed researcher (without a doctorate) and with only two years of professional experience, albeit from an eco-municipality (Ølen).

The NAMIT project began the same year as the MIK program—two years before I started at WNRI—and continued until 1992. Petter Næss at the Norwegian Institute for Urban and Regional Research (NIBR) led NAMIT, carried out in collaboration with WNRI, the Institute of Transport Economics, the Department of Landscape Architecture at the Norwegian University of Agriculture (later the University of Life Sciences), and the Centre for Development and the Environment at the

[12] Jansen (1991).

[13] Aall et al. (2002).

University of Oslo. Petter Næss was a close friend of Karl Georg Høyer, and they maintained a strong professional collaboration throughout their research careers. I also had the pleasure of working extensively with Petter Næss. Unlike the Eco-Municipality Programme, the NAMIT project was not directly involved in local political processes. It therefore falls under the category of research "on" rather than research "for" municipalities, making it more of a theoretical feasibility study compared to the Eco-Municipality Programme, which began two years after NAMIT. Many of the perspectives on the scope of municipal action for sustainable development that emerged within NAMIT were later tested in "real life" through the Eco-Municipality Programme.

NAMIT analyzed alternatives for land use and development patterns in three Norwegian urban areas: Horten, Sogndal center, and Malvik (in the eastern part of Trondheim). In each study area, researchers constructed an environmentally prioritized and a trend-based development alternative, which were assessed against the objective of sustainable development. The analyses were based on a detailed and comprehensive operationalization of the sustainability goal, structured as a goal hierarchy consisting of seven main objectives and 32 sub-objectives. This hierarchy was also developed by the researchers. The first two main objectives address global environmental issues, the next three focus on more conventional local environmental problems, and the final two concern economic and social aspects:

1. Minimize energy consumption and emissions of global/national significance.
2. Preserve biological resources.
3. Minimize local noise and pollution.
4. Provide good opportunities for outdoor recreation.
5. Preserve landscape and cultural values.
6. Be economically viable for society.
7. Contribute to achieving welfare and social objectives.

NAMIT drew two main conclusions. First, significant benefits for nature and the environment can be achieved while simultaneously meeting social and economic objectives, by concentrating spatial development, promoting densification in areas where technical interventions have already occurred, and utilizing each building plot more efficiently. This approach contrasts with the land-intensive and dispersed development pattern that characterized much actual urban growth in Norway and most other Western countries at the time. In particular, environmental gains can be substantial if a development strategy that reduces transport distances and conserves land is combined with strengthened public transportation and measures to limit car traffic in urban centers.[14]

Second, the NAMIT project illustrates that it is naive to believe that an ambitious national environmental policy will almost automatically trigger a correspondingly ambitious local environmental policy. The gap between the environmental and trend scenarios, together with the project's own surveys of residents, elected officials,

[14] Næss (1992).

and municipal administration, indicates another obstacle: insufficient local political support for environmentally friendly development alternatives. Furthermore, the surveys showed that the gap between attitudes and an environmentally friendly ideal is greater among residents than among elected officials and municipal employees (see Footnote 14).

My first major assignment at WNRI was to lead the Eco-Municipality Programme, initially as a pilot project in 1990 and subsequently as a six-year main project from 1991 to 1996.

In Norway, the idea of securing state funding for a national eco-municipality program was first introduced to a selection of municipalities by the organization Norwegian Forum for Ecological Economics in the early 1980s.[15] The initiative failed, primarily due to skepticism from both the municipalities and, in particular, the Ministry of the Environment. Both were concerned about those aspects of the original idea that involved introducing a local barter economy.[16] In 1987, the municipality of Sogndal, represented by Liberal Party mayor Bjørg Rimeslaatten on behalf of ten self-declared eco-municipalities, contacted Karl Georg Høyer. They requested professional assistance from WNRI to prepare an application to the Ministry of the Environment for a preliminary project designed to pave the way for a full Eco-Municipality Programme, in which pilot municipalities set out to independently develop strategies to follow up on key recommendations from the Brundtland Report.[17] WNRI was awarded the pilot project, and I was hired in June 1990 to carry it out.

Municipalities wishing to participate further in a main project were required to formally adopt an eco-municipality declaration. This statement, developed in collaboration with the municipalities during the pilot project, included seven points:

1. Integrate ecological objectives into municipal planning.
2. Environmental impacts must be assessed in municipal case processing.
3. Mobilize the public to utilize local knowledge and natural resources to promote sustainable development.
4. Ensure conservation and sustainable use of local natural and cultural heritages.
5. Provide local contributions to solving global environmental problems.
6. Address local environmental protection challenges.
7. Emphasize strong local governance and engagement.

The pilot project led to a full scale research and development program funded by the Ministry of the Environment and the KS. Nine rural municipalities participated: Forsand, Lardal, Nore og Uvdal, Nesseby, Sogndal, Suldal, Tingvoll, Tolga, and Ølen. One municipality—Tysvær—withdrew during the pilot project, citing concerns that the eco-municipality profile would conflict with its efforts to further develop what eventually became Europe's largest facility for processing gas and light oil: the Kårstø facility.[18]

---

[15] Myrhaug (1987).

[16] Aall (1997).

[17] Høyer (1987).

[18] Aall (1991).

When Nine Eco-Municipalities, Instead of 91 MIK Municipalities, Became Norway's Face of Sustainability at a World Congress in Oslo in 1991

In 1991, the International Union of Local Authorities (IULA) organized its 30th world congress. The venue was Oslo, the theme of the year was sustainable development, and the then Prime Minister of Norway—and former chair of the UN Commission on Environment and Development—Gro Harlem Brundtland, delivered the opening address. The congress was set to discuss and adopt the so-called Oslo Declaration on the responsibility of municipalities in the work toward sustainable development.

An important part of the conference was an exhibition intended to showcase how municipalities around the world began working toward sustainable development. The Ministry of the Environment was invited to present the MIK program and its announced reform; however, for reasons unknown, they declined the invitation. Perhaps the event was considered too "political" or too focused on "municipalities" for the ministry to participate.

The Norwegian organizers were somewhat dismayed. It would have been embarrassing if the host country had nothing to present, so the organizers contacted the Idébanken Foundation. The foundation, in turn, reached out to us and asked if we could help create a Norwegian contribution to the exhibition focused on the Eco-Municipality Programme. The foundation was part of the Alternative Future Study. Its purpose was to collect and document practical examples of alternatives to the dominant model of societal development, which assumed perpetual economic growth. Thus, it came to pass that the small Eco-Municipality Programme—not its "big brother" the MIK programme—became Norway's face to the world at the IULA exhibition. That conference was meant to set the tone for how the world's municipalities would contribute to achieving sustainable development.

The collaboration with the Idébanken Foundation continued in the years that followed, in connection with Local Agenda 21 and efforts to carve out local scope for action in consumption-oriented environmental and climate policy. The collaboration also led to the foundation's key professional, John Hille, receiving several project assignments at WNRI. I eventually developed a close relationship with John.

While the Ministry of the Environment, in the first and so far, only white paper on municipal environmental protection from 1991,[19] describes the MIK as a "development programme to strengthen municipal environmental management" (p. 12). Without any explicit reference to the goal of sustainable development, the ministry refers to the Eco-Municipality Programme as follows (p. 14):

> A development-oriented municipal initiative that seeks to leverage local resources, local engagement, and local governance in accordance with sustainable development.

[19] Miljøverndepartementet (1991).

The Eco-Municipality Programme consisted of four projects:

- Ecologically based municipal planning.
- Municipal management of cultural landscapes.
- Environmental knowledge training for policymakers.
- Municipal environmental auditing.

The first two projects involved a series of meetings held throughout the municipalities. We, as researchers, prepared technical briefs in advance and presented them at the meetings for representatives from the administration and municipal council. Attendees then discussed and further developed the material. Site visits were often included, during which representatives from the host municipality showcased examples of issues or solutions related to the topics and challenges selected for each meeting. The meetings were summarized in dedicated minutes.

The political training program was conducted within each municipality. It consisted of sessions for the politicians, often organized as an hour-long segment before or after a regular municipal council meeting. The purpose of these sessions was to enhance the politicians' knowledge of current sustainability topics. After the training program, participating politicians reported that they had changed their attitudes on environmental issues and had also altered their voting behavior in certain cases (see Footnote 16).

The most comprehensive project—the Municipal Environmental Audit—was funded independently of the Eco-Municipality Programme through a dedicated research program under the Research Council of Norway. It included only some of the eco-municipalities, plus several other municipalities. This was the first project in Norway to place formalized systems for environmental management on the agenda in a municipal context. We developed and tested a municipality-adapted method for environmental auditing, partially inspired by the EU's Eco-Management and Audit Scheme (EMAS), which was introduced in 1995. The adaptation consisted of combining field-based environmental inventories with conventional methods of management auditing. The results were published by Kommuneforlaget as a documentation report[20] and a handbook[21] and distributed free of charge to all Norwegian municipalities.

We Will Never, Ever, Ever Do This Again!

In the project "Municipal Environmental Audit", we tried a new methodological approach. We borrowed the logic from the laboratory—where variables are isolated and processes are observed under controlled conditions—and transferred it into an organization. That's a practice often referred to as *sociotechnical experimentation.* In one of the project municipalities, Steinkjer, we likely stretched this methodological approach too far by adding yet another element:

[20] Aall and Sæther (1996).

[21] Aall et al. (1996).

observers. In the methodological literature, this is known as the "Fishbowl Method."

The idea behind environmental auditing is simple. You determine the extent to which a given environmental objective has been achieved. If not, you identify the reasons why. The audit investigation consists of document analysis, often supplemented by site inspections, and—most importantly—interviews with individuals who have, or should have, insight into the work.

In Steinkjer, we chose to conduct the experiment on a project to convert a former military camp into "the country's most environmentally friendly school building." That was the municipality's stated goal. After the building was completed, it turned out to be far less environmentally friendly than intended, and we set out to find out why.

Based on a review of various documents, we suspected that the building committee had made a decision at some point during the construction process that weakened the environmental ambitions. To shed light on this, we decided to conduct an audit interview with the building committee, but we also decided to do it as a public event. We invited about a dozen observers: representatives from national environmental authorities, regional administration, and municipal politicians. We then placed the three members of the municipal building committee on a stage, and we researchers (there were two of us) acted as auditors, conducting the interviews in full public view.

We discovered that, indeed, a critical decision had been made. The committee had agreed to a proposal from the chair to lower the environmental ambitions in order to save money. This was also done without informing the project's client, the municipal council. When the chair confirmed this from the stage, tears began to flow.

After that session, Steinkjer municipality decided they would never, ever, ever use the environmental audit tool in their environmental work again—and we gained one more experience in how *not* to conduct a development project.

The municipalities aspired to achieve changes toward sustainability through their participation in the program. However, the evaluation's main conclusion was that the initiative never truly "took off."[22] The reasons were partly related to characteristics of the municipalities themselves, and partly to the organization and prioritization of activities within the program. At the same time, the evaluation pointed out many interesting traces of the eco-municipality concept in the municipalities' activities during the program period. In contrast to the general findings from the MIK program evaluation, the eco-municipalities showed several examples of using relatively strong planning instruments in municipal environmental policy—for instance,

[22] Ramsdal (1997).

to ensure the protection of cultural landscapes. There were also examples of eco-municipalities implementing measures to reduce greenhouse gas emissions, protect biological diversity, and promote global equity.

The Eco-Municipality Tingvoll Breaks the Law

One of the eco-municipalities, Tingvoll, was the first in Norway to introduce home composting as part of the municipal waste management system. Local enthusiasts had also started producing home compost bins. The bin was named after a former fellow student of mine at the Agricultural University of Norway, who had designed it: Knut Haga. It became known as the "Haga compensate". The Haga compensate was a great success and was sold to many Norwegian municipalities. An educational version with a glass wall was also developed so that you could watch the composting process. This version was sold to kindergartens across the country, including the one my children attended (Henjahaugane Kindergarten in the former Leikanger municipality).

Back when Tingvoll began its pioneering work with home composting, national health authorities and environmental authorities were at odds. Around 1990, a regulation authorized by the old Public Health Act of 1860 initially banned home composting in densely populated areas without special permission from the local health board. The health authorities were worried about infection risks and hygiene. They were also concerned about attracting pests (especially rats) and odor problems. Tingvoll municipality chose to ignore this conservative interpretation. They introduced home composting as an active measure to reduce the amount of waste that needed to be transported and landfilled.

The environmental authorities were aware of Tingvoll municipality's expertise in this area. When other municipalities contacted the authorities for advice, the case officers were formally required to refer to the health authorities' prohibition, but at the same time they would advise callers to contact the technical department in Tingvoll municipality for useful guidance and tips.

What happened in Tingvoll (and in a few other municipalities, such as Fredrikstad) forced a change in national regulations. The experiences from Tingvoll showed that if you used eco-labeled, insulated, and sealed compost bins—like the one Knut Haga developed—the hygienic challenges were minimal. As a direct result of these local acts of civil disobedience, the regulations were amended. It then became possible for municipalities to grant reductions in waste collection fees to residents who composted at home, provided they completed a training course and used approved equipment.

From State-Initiated Program to State Reform

The MIK program was originally intended to last three years. However, after just over two years—in autumn 1990—the government decided to continue the program as part of a reform. This decision came through the first and only white paper to date submitted to the Storting on municipal environmental protection (see Footnote 19).

The same year, in June 1990, I began working at WNRI, where my role was to establish a research field on municipal environmental policy under the leadership of the then director of WNRI—and later a very good friend—Karl Georg Høyer. Initially, we were a core group of three researchers: Karl Georg, Kyrre Groven, and myself. Over time, the group expanded with additional researchers. As part of this development, an important and long-term collaboration began in 1995 with Professor William M. Lafferty and ProSus, which he led.

The enthusiasm generated by the World Commission's report and the launch of the sustainable development goal also influenced the MIK and its subsequent reform. However, the government white paper on municipal environmental protection reflects very little of this enthusiasm. Beyond a few general references to the sustainability goal (a total of ten), the white paper mentions only once any concrete and specific state expectations regarding municipalities' contributions to sustainability efforts—and then only in relation to nature management.

No comprehensive and coordinated evaluation of the MIK reform was conducted, unlike the evaluation carried out for the MIK program. Instead, researchers conducted several limited and largely uncoordinated questionnaire-based surveys by NIBR,[23] the KS,[24] and ProSus in collaboration with WNRI.[25]

That We Should Have a Full-Time Employee Working Exclusively on Environmental Issues in My Small Village? No, That's Unthinkable!

In 1995, as part of the Environmental Audit Project, I organized a study trip to England. We had become fascinated and impressed by the English systems for environmental management and their attempts to transfer this systematic approach to environmental protection from private companies to municipalities. Representatives from eight Norwegian municipalities participated in the study tour: Bærum, Fredrikstad, Rakkestad, Sogndal, Stavanger, Steinkjer, and Tingvoll. All had a full-time environmental manager funded through the MIK reform.

We were taken to several municipalities, all of which were large by Norwegian standards in terms of population, and where the office facilities appeared somewhat "worn," also by Norwegian standards. This was only a few years after the Thatcher era, which had resulted in significant cuts to public—and

[23] Hovik and Johnsen (1994).

[24] KS (Unpublished). Cited in Aall et al. (2001).

[25] Lafferty et al. (1998).

especially municipal—finances. Yet their environmental management systems were impressive.

Our hosts, quite reasonably, wanted to hear how Norwegian municipalities worked with environmental protection. We explained how little progress we had made in implementing environmental management systems. Then, almost in passing, we mentioned that nearly all Norwegian municipalities at the time had a full-time environmental manager. We also mentioned—again, almost in passing—the population sizes of the municipalities participating in the study tour, with Tingvoll being the smallest at 3000 inhabitants.

During one of the visits, in Newcastle, I was in the middle of explaining how poorly Norway was doing on environmental management. One of the hosts interrupted me, saying, "Wait, are you saying that Tingvoll municipality, with 3000 inhabitants, has a full-time position dedicated exclusively to environmental issues?" I confirmed this, whereupon he almost muttered to himself, "That we should have a full-time employee working exclusively on environmental issues in my small village? No, that's unthinkable!" He was evidently from a village on the outskirts of Newcastle itself.

Based on a distinction between concentrated and dispersed problem generation and concentrated and dispersed distribution of effects, researchers at NIBR developed a typology of four types of environmental problems.[26] *Local environmental problems* are cases where both the cause and its effects are confined to the local area. *Dispersion problems* occur when local emissions or other negative local environmental impacts also have effects beyond municipal boundaries. *Accumulation problems* arise when the local community "imports" the effects of environmental problems—for example, local fish mortality because of acid rain. In cases where both the generation of the problem and its effects are dispersed—for example, CO_2 emissions from private car use contributing to the greenhouse effect—the typology refers to such issues as *commons problems.*

Based on the results from NIBR's national survey, Hovik and Johnsen (see Footnote 23) found that Norwegian municipalities generally appear to prioritize local environmental issues. According to their data, municipal contributions to addressing problems of a more global nature—such as climate change and biodiversity loss — were rarely prioritized. Naustdalslid and Hovik (see Footnote 26) argue that environmental problems of a global character provide little reason to expect local motivation to take responsibility for solving them. These collective problems are created by many different activities across a large area and affect everyone living there. These characteristics suggest that there will be little local political activity to address them. On the contrary, they create a motivation to free ride on the efforts of others.

Despite the view that local efforts should focus on local environmental issues, Naustdalslid and Hovik (see Footnote 26) acknowledge that municipalities may also

[26] Naustdalslid and Hovik (1994).

have a legitimate role in addressing distinctly global environmental challenges. They identify two approaches for modifying institutional frameworks to encourage municipalities to give higher priority to global environmental issues. The first involves local planning, which presupposes a shift in value priorities among the population. The second entails changes to national framework conditions for municipal governance. How these two perspectives manifested in practice would soon become evident through in-depth case studies of pioneering municipalities—offering a complementary lens to the broad-based surveys conducted by NIBR.

Local Agenda 21 Arrives in Norway—Eventually!

In 1992, the United Nations Conference on Environment and Development in Rio de Janeiro adopted the action plan for sustainable development: Agenda 21. Chapter 28, often referred to as Local Agenda 21 (LA21), outlines the role of local authorities. The chapter itself does not specify the content of a municipal policy for sustainable development (it comprises only three out of 700 pages in the Agenda 21 document). Instead, the chapter refers to the main text so that local authorities are, in principle, faced with the same challenges and requirements as national authorities.

At the outset of what I have termed the consolidation period in municipal environmental policy, two competing approaches emerge regarding the content of municipal environmental policy: one "local" and one "global." The "local" approach is associated with the MIK reform. The Ministry of the Environment adheres to this local approach, which is linked to the continuation of the MIK reform. The environmental movement—particularly the Idébanken Foundation—advocates for promoting Local Agenda 21 as something distinct and more global than the MIK approach. Over time, the KS has also aligned itself with the "global" approach. WNRI and ProSus shared this perspective, whereas other research communities—notably researchers at NIBR—tended to favor the "local" approach.

Interestingly, the first circular that provides a detailed description of governmental expectations for municipal environmental policy, from 1993, opens with the following sentence (p. 1)[27]:

> The responsibility of municipalities to address global environmental and development challenges was established in the Brundtland Commission's report from 1987.

This sentence is interesting for several reasons. First, the World Commission's report *does not* address municipalities and therefore *does not* mention anything about municipal responsibilities. According to the UN report, the sustainability project was a *national* and *intergovernmental* political responsibility. Second, the only paragraph in the entire report that mentions municipalities points out that their role should typically concern local environmental issues. Third, the circular must be understood as part of Norway's follow-up to Agenda 21; however, Chap. 28 of Agenda 21 was

[27] Miljøverndepartementet (1993).

not included in the original draft. It was added as a result of lobbying by representatives of local authorities, primarily through pressure from the ICLEI, which is a subgroup of the IULA.[28] When the Ministry of the Environment eventually chose to include the role of municipalities in relation to global environmental issues, it appeared to be a consequence of the municipalities themselves seeking such a role. This was not based on what the World Commission's report stated.

The Country's First Local Agenda 21 Was Adopted by Sogndal Municipality in 1996. Coincidence? Hardly!

In 1995, after I had invited myself to a meeting with Professor William M. Lafferty in Oslo—among other things, to discuss the Eco-Municipality Programme and our work on municipal environmental auditing—I invited him to visit Sogndal and WNRI to inform us researchers about ProSus and its work with Local Agenda 21.

Once the internal seminar was underway, Karl Georg Høyer pointed something out to me. The Sogndal municipal council would be considering its Environmental and Natural Resource Programme later that day. (Karl Georg was also an active participant in local politics for the Socialist Left Party.) He suggested we contact the municipal chief executive and offer Bill—introduced as one of the world's leading experts on Local Agenda 21—to give a presentation to the council in connection with their deliberations.

No sooner said than done.

William M. Lafferty had time to take a quick look at the municipality's Environmental and Natural Resource Programme before we went down to the town hall. Then, true to his American improvisational style, he began his introduction by asking whether the Sogndal municipal council would consider renaming the plan "Local Agenda 21." He pointed out that in both form and content, the document already met all the requirements of a Local Agenda 21. And then he added, "Then you will become Norway's first municipality to adopt a Local Agenda 21."

Which is exactly what the council did.

It was not until the end of 1996, four years after the adoption of Agenda 21, that policy signals emerged indicating that the Ministry of the Environment was beginning to abandon its assessment that the MIK reform should be regarded as a sufficient follow-up to Chap. 28 of Agenda 21. In a joint letter dated December 9, 1996, the Ministry of the Environment and the KS informed municipalities, county authorities, and county governors of a forthcoming transition from "MIK" to "LA21" as the framework for municipal environmental policy. The letter commented on this transition as follows:

[28] Hams (1994).

> Although municipalities are already making significant environmental efforts, we are facing new and major challenges in steering development towards a more environmentally friendly and sustainable direction. The following points are central:
>
> - We must see a clear and concrete shift towards sustainable development.
> - The global perspective of local environmental efforts must be recognized and made visible.
> - Local Agenda 21 should be a process involving the entire local community, with particular emphasis on increasing the participation of women and children.
> - The work should have a long-term (multi-generational) perspective.
> - The objective of improved quality of life must be integrated so that both health issues and the fundamental economic and social consequences of the transition to sustainable development are made visible.

The following spring, in June 1997, the Ministry of the Environment presented White Paper 58 (1996–1997), *Environmental Policy for Sustainable Development*. In it, LA21 appears as one of four main priority areas. However, although the white paper clearly states that LA21 will now be the focus, it is less clear about what the global environmental challenges entail for municipalities. The Parliament's Energy and Environment Committee addressed this ambiguity in its recommendation on the white paper. The committee took the opportunity to emphasize that municipalities also have a responsibility with respect to global environmental issues (p. 16)[29]:

> Local Agenda 21 must be a municipal environmental plan that provides guidance on how the local community can reduce its environmental impact both globally and regionally, and how it can contribute to a more equitable distribution of resources worldwide.

An important follow-up to the signal in White Paper 58 regarding the commitment to LA21 took place from February 9–11, 1998. Seven hundred participants from 150 municipalities gathered in Fredrikstad for a national conference on LA21. The event became something of a milestone in Norwegian environmental policy. It later stood out as the starting point for widespread support for LA21 initiatives in Norwegian municipalities. The meeting adopted the Fredrikstad Declaration, with two main objectives:

> Sustainable societal development that ensures quality of life and the basis for existence both today and for future generations; and that activities in our local communities take place within the carrying capacity of nature, both locally and globally, and that we therefore seek to reduce natural resource consumption and environmental impact. The declaration identified the following five points as "the greatest challenges in a Norwegian local Agenda 21 process": Reduce consumption (including energy consumption); develop more sustainable transportation; address climate issues sustainably; preserve biological diversity; and develop a sustainable local economic policy.

The main message of the Fredrikstad Declaration is that municipalities must now take responsibility for initiating LA21. The declaration urged municipalities to endorse it by the end of 1998. Eventually, 270 municipalities, all 19 county authorities, and several voluntary organizations adopted the declaration.[30] This represents a markedly

---

[29] Energi-og miljøkomiteen (1998).

[30] Aall et al. (2006).

different approach from the prevailing state governance signals to Norwegian municipalities. It is more in line with the recommendations that emerged from the NAMIT project and the Eco-Municipality Programme a decade earlier.

The UN has conducted three comprehensive reviews of progress in implementing Agenda 21 since 1992: Rio + 5 in New York in 1997, Rio + 10 in Johannesburg in 2002, and Rio + 20 in Rio de Janeiro in 2012. In connection with Rio + 5, I participated in the first European comparative study of LA21 initiatives, led by William M. Lafferty.[31] One conclusion was that Chap. 28 on LA21 was the part of Agenda 21 that received the greatest support and had the most significant impact on practical policy. Further success of this work was largely limited by the absence or lack of effectiveness of national Agenda 21 strategies. Compared with seven other countries (Finland, Sweden, Germany, Austria, the Netherlands, the United Kingdom, and Ireland), Norway started relatively late; however, as acceptance grew that LA21 was something "more" and "different" than the MIK reform, Norway assumed a leading position among the eight countries included in the study.

Where Do Norwegian Municipalities Stand Today in Their Efforts to Promote Sustainable Development?

I began the chapter by introducing the theory of political cleavage and presented a historical division of municipal environmental work: the pre-institutional phase, the experimental period (1985–1992), the reform period (1992–1997), and the consolidation period (after 1997). However, developments after 2002 have been less uniform than in the preceding periods. What we observe following the enthusiastic era of the MIK reform and Local Agenda 21 is that much of the local "counter-politics" has subsided. Local environmental policy has increasingly been transformed from an arena for innovation and system-critical opposition into a more conformist structure for implementing state directives.

This development reflects the fundamental divide I pointed out at the outset—the tension between municipal autonomy and national governance—but in a new and more fragmented form. Whereas the MIK reform and the early eco-municipalities were based on a generalist model with a mandate to think holistically across sectors, the developments of recent decades have fragmented environmental responsibility and weakened the political force of the broad sustainability objective. To understand how the scope for local action has been altered and at times narrowed, we must examine three parallel processes more closely: deinstitutionalization, dilution, and specialization.

[31] Lafferty and Eckerberg (1998).

Deinstitutionalization of Municipal Environmental Protection Efforts

Since the 2000s, a form of *deinstitutionalization* of municipal environmental policy has taken place. We see this in the reduction of administrative capacity in the environmental sector that had been built up during the 1990s in many municipalities, particularly in small and medium-sized rural municipalities. After peaking in 1994, when approximately 95% of municipalities had appointed an environmental officer, the proportion declined dramatically. The cause was the discontinuation of earmarked state transfers for financing such positions in 1997. From 1996 to 2000, 27% of municipalities eliminated the position of municipal environmental officer, while 31% either reduced or merged the position with other functions. By that time, only 28% of municipalities (91 municipalities, or 20% of all municipalities) had appointed a full-time employee dedicated exclusively to environmental issues.[32] This is the same number of municipalities (though not the same ones) that participated in the MIK program.

Equally important as the changes in the number of municipal environmental officers are the shifts in national policy directives. From 2002, all forms of state support for LA21 effectively ceased. Over time, state expectations for municipal environmental policy also shifted. The main objective of the MIK reform was to build administrative capacity and political competence within municipalities to handle new state-assigned tasks in the environmental sector. In 2001 and 2002, an extensive discussion took place about delegating new environmental responsibilities to municipalities. In the end, the Ministry of the Environment transferred a limited number of distinctly local environmental tasks to the municipalities. Gone were the expectations for municipal efforts in, for example, energy and climate policy. The transfer of responsibilities related to biodiversity was reduced to two tasks: mapping rare species and constructing deer fences (see Footnote 30).

Several times over the years in political debates—particularly during parliamentary and municipal elections—the question has been raised about reintroducing earmarked transfers to strengthen municipalities' administrative capacity in the environmental sector. In recent years, this has been linked to municipalities' increasing involvement in climate-related issues. However, these proposals have not gained traction. Currently, no studies document municipal administrative capacity in the environmental sector with the same precision as the studies conducted during and after the MIK reform. Nevertheless, all studies from the past decade that examine barriers and obstacles reported by municipalities in their work with environment, sustainability, climate, or other sustainability-related tasks identify insufficient administrative capacity as the primary obstacle.[33]

[32] Bjørnæs (2006).

[33] Vindegg et al. (2022).

The Dilution of the Sustainability Goal Also Occurs at the Local Level

As demonstrated in the chapter reviewing the evolution of the sustainability concept, a *dilution* and subsequent *distortion* of the sustainability objective have occurred in both Norwegian and international discourse since its introduction in 1987. This development is also evident in perspectives on local environmental protection efforts.

Chapter 28 of Agenda 21 served as the 1990s "roadmap" for local sustainability efforts. It effectively established a single objective for municipalities: through an LA21 process, to address all topics and goals in Agenda 21 that the municipalities deemed locally relevant. By 2015, this had shifted. One of the 17 main Sustainable Development Goals (SDGs) is SDG 11, Sustainable Cities and Communities (SDG 11, Sustainable Cities for short). In principle, the goal should apply to all local communities, but the focus is on large cities. The ten sub-targets address distinctly urban issues, such as "affordable housing," "affordable and sustainable transport systems" (with an emphasis on "affordable" in the explanation of the goal's content), and "inclusive and sustainable urbanization." The social and economic dimensions of sustainability are predominant. Only one sub-target—Target 11.6—includes the word *environment* in its title: "Reduce environmental impacts of cities."

The prioritization reflected in the sub-goals of SDG 11 is naturally linked to the rapid urbanization occurring globally (60% of the world's population is expected to live in cities by 2030) and the significant social challenges faced by many large cities, as illustrated by the fact that urban slum areas currently house approximately 1 billion people.[34]

This narrowing of municipal sustainability policy to focus solely on major cities, emphasizing social and economic challenges at the expense of environmental concerns, has led municipalities internationally—and in Norway—to also address many of the remaining 16 SDGs. A guide published by the Global Task Force of Local and Regional Governments (a collaboration among the leading international networks for local authorities) on local implementation of the UN SDGs is titled *Roadmap for Localizing the SDGs: Implementation and Monitoring at Subnational Level*. The guide defines *localizing* as "the process of taking into account subnational contexts in the achievement of the 2030 Agenda" (p. 6).[35] The result of this approach has therefore been an expansion—not a narrowing—and, as argued in the chapter on the broadening of the sustainability concept, a dilution of its conceptual content.

A 2019 report by the consultancy firm Asplan Viak on behalf of the KS confirms that many Norwegian municipalities choose to address the entire set of UN SDGs, not just SDG 11. The report notes that this is being done not only within what was previously framed as "environmental policy," but increasingly under the broader heading of "community planning."[36] The report points out that for many, this is

---

[34] Groven and Aall (2020).

[35] Global Taskforce of Local and Regional Governments, UN-Habitat, and the United Nations Development Programme (UNDP) (2016).

[36] Bang-Andersen et al. (2019).

linked to the fact that the municipal sector is increasingly facing challenges related to demographic factors (such as population decline, a growing proportion of elderly, and gender imbalances) and widening disparities in living conditions, and that these challenges are competing with environmental issues for attention.

At the same time, the report highlights the increasing expectations placed on the sector as a service provider while financial constraints are tightening. It asserts that the SDGs are therefore perceived as relevant and useful for the municipal sector's efforts to provide residents with the best possible services, promote community development, and fulfill its role as an employer within increasingly limited budgets. Thus, there is less emphasis on the environmental aspects of the sustainability goals and more focus on what has gradually come to be characterized as social and economic sustainability. The following year, Nordlandsforskning published a study that confirmed the conclusions of the Asplan Viak report: municipalities report that working with the UN SDGs has strengthened their efforts in integrated community planning.[37]

The efforts to promote sustainable development have thus evolved from serving as an incentive for municipalities to engage in addressing not only local but also global environmental challenges, and to contribute to a more equitable distribution between poor and rich countries. They have become a stimulus for comprehensive societal planning that extends far beyond the original 1990s understanding of what constitutes sustainability work.

Specialization of Local Sustainability Efforts

The third developmental trend evident in municipal environmental policy is a specialization—or perhaps more accurately, a narrowing. This represents the most distinct change since the turn of the millennium. At the beginning of the reform period, the terms used were "environment and natural resource program" and "environmental protection officer." At the start of the consolidation period, the terms were "Local Agenda 21 plan" and "sustainability advisor." The terminology now most used is "climate plan" and "climate coordinator" for municipal plans and personnel specifically working with environmental issues in Norwegian municipalities. In the 2010s, a range of national policy directives, guidance materials, and financial support schemes were developed that focused specifically on climate. In the chapter on climate policy, I provide a detailed account of experiences stemming from the significant surge in municipal climate policy that occurred throughout the 2000s.

NAMIT Revisited 30 Years Later

[37] Lundberg et al. (2020).

In 2018, a revisited NAMIT project was launched under the title Sustainable Urbanisation Requirements of Small and Medium-Sized Urban Settlements and Their Surroundings (Surround). This time, the project was led by WNRI, with partner institutions different from those in NAMIT, although many of the same individuals took part. Not the least of these was the leader of the original NAMIT project, Petter Næss, who later became a professor at the Norwegian University of Life Sciences (NMBU). One of the members of the original research group on municipal environmental protection at WNRI—now Dr. Kyrre Groven—led the Surround project. The same case areas as in NAMIT were included, plus a doctoral dissertation.

From the press release on the occasion of colleague Karen Richardsen Moberg's defence, I quote the following:

"A radical transformation of land-use planning is needed... We must bring back tools that have gathered dust in the toolbox since the neoliberal wave of the 1980s and 1990s and find alternatives to growth if we are to succeed in reducing the loss of nature and species and lessen the burden on the climate." (see Footnote 2). Truly: revisited!

References

Aall, C. (1988). *Miljøårboka* [The Environmental Yearbook]. Samlaget.

Aall, C. (1991). *Økokommuneprogram for 9 norske kommunar* [Eco-municipality program for 9 Norwegian municipalities] (VF-report 14/91). Western Norway Research Institute.

Aall, C. (1997). *Det kommunale miljøvernet—fra lokalt til globalt? Oppsummering av det norske økokommuneprogrammet 1989–96* [Municipal environmental protection—from local to global?] (VF-report 2/97). Western Norway Research Institute.

Aall, C. (2000). *Forandring som forandrer? Fra miljøpolitikk til bærekraftig utvikling i norske kommuner* [Change that changes? From environmental policy to sustainable development in Norwegian municipalities] [Doctoral dissertation, Aalborg University]. Western Norway Research Institute.

Aall, C. (2002). Framveksten av det kommunale miljøvernet i Norge. Fra miljøvern til bærekraftig utvikling? [The emergence of municipal environmental protection in Norway]. In C. Aall, K. G. Høyer, & W. M. Lafferty (Eds.), *Fra miljøvern til bærekraftig utvikling i kommunene. Erfaringer med Lokal Agenda 21* (pp. 29–70). Gyldendal akademisk.

Aall, C., Erstad, E., & Vestby, S. E. (1996). *Håndbok i kommunal miljørevisjon* [Handbook of municipal environmental auditing]. Kommuneforlaget.

Aall, C., Høyer, K. G., & Lafferty, W. M. (Eds.). (2002). *Fra miljøvern til bærekraftig utvikling i kommunene. Erfaringer med Lokal Agenda 21* [From environmental protection to sustainable development in the municipalities]. Gyldendal akademisk.

Aall, C., Lafferty, W., & Lindseth, G. (2001). *Ansvars- og myndighetsfordeling mellom stat og kommune på miljøområdet* [Distribution of responsibility and authority between state and municipality] (VF-report 7/01). Western Norway Research Institute.

Aall, C., Lindseth, G., & Lafferty, W. M. (2006). Lokal Agenda 21 i Norge. In W. M. Lafferty, C. Aall, G. Lindseth, & I. T. Norland (Eds.), *Lokal Agenda 21 i Norge* (pp. 239–258). Unipub.

Aall, C., & Sæther, B. (1996). *Når miljømål møter den kommunale virkeligheten* [When environmental goals meet municipal reality]. Kommuneforlaget.

Bang-Andersen, S., Plathe, E., & Hernes, M. B. (2019). *Prioriterte mål i kommunalt og fylkeskommunalt planarbeid* [Priority goals in municipal and regional planning] (Asplan Viak Report). Asplan Viak.

Bjørnæs, T. (2006). Lokal Agenda i Norge som lokalt fenomen [Local Agenda in Norway as a local phenomenon]. In W. M. Lafferty, C. Aall, G. Lindseth, & I. T. Norland (Eds.), *Lokal Agenda 21 i Norge* (pp. 65–93). Unipub.

Deegan-Krause, K. (2007). New dimensions of political cleavage. In R. J. Dalton & H. D. Klingemann (Eds.), *The Oxford handbook of political behavior* (pp. 538–556). Oxford University Press.

Energi-og miljøkomiteen. (1998). *Innstilling fra energi- og miljøkomiteen om miljøvernpolitikk for en bærekraftig utvikling—dugnad for framtida* [Recommendation from the Standing Committee on Energy and the Environment on environmental protection policy for sustainable development] (Innst. S. nr. 150 (1997–98)). Stortinget.

Global Taskforce of Local and Regional Governments, UN-Habitat, & the United Nations Development Programme (UNDP). (2016). *Roadmap for localizing the SDGs: Implementation and monitoring at subnational level.* Global Taskforce of Local and Regional Governments.

Groven, K., & Aall, C. (2020). *Korleis kan distriktskommunar arbeide med berekraftig utvikling?* [How can rural municipalities work with sustainable development?] (VF-notat 3/2020). Western Norway Research Institute.

Hams, T. (1994). Local environmental policies and strategies after Rio. In J. Agyeman & B. Evans (Eds.), *Local environmental policies and strategies.* Longman.

Hovik, S., & Johnsen, V. (1994). *Fra forsøk til reform: Evaluering av MIK-programmet* [From experiment to reform: Evaluation of the MIK program] (NIBR Rapport 1994:23). NIBR.

Høyer, K. G. (1987). Økokommunar—prosjektnotat [Eco-municipalities - project note]. In E. Myrhaug (Ed.), *Økokommunen—tenke globalt, handle lokalt.* Venstres Opplysningsforbund.

Høyer, K. G., & Selstad, T. (1991). *Regionalpolitikkens økologiske grunnlag* [The ecological basis of regional policy]. Western Norway Research Institute/NordREFO.

Jansen, A. (1991). *Reform og resultater. Evaluering av forsøksprogrammet Miljøvern i kommunene* [Reform and results. Evaluation of the MIK program]. NAVF.

Kvaløy, S. (1979). Økopolitikk og økofilosofi [Ecopolitics and ecophilosophy]. In S. Kværner & J. Rattsø (Eds.), *Økopolitikk og populisme.* Pax.

Lafferty, W., Aall, C., & Seippel, Ø. (1998). *Fra miljøvern til bærekraftig utvikling i norske kommuner* [From environmental protection to sustainable development in Norwegian municipalities] (Report 1998:2). ProSus.

Lafferty, W. M., & Eckerberg, K. (Eds.). (1998). *From the Earth Summit to Local Agenda 21: Working towards sustainable development.* Earthscan.

Lindseth, G. (2004). The Cities for Climate Protection Campaign (CCPC) and the framing of local climate policy. *Local Environment, 9*(4), 325–336. https://doi.org/10.1080/1354983042000246252

Lipset, S. M., & Rokkan, S. (Eds.). (1967). *Party systems and voter alignments: Cross-national perspectives.* Collier-Macmillan.

Lundberg, A. K. A., Bardal, K., Vangelsten, B. V., Reinar, M. B., Bjørkan, M., & Richardson, T. (2020). *Strekk i laget: En kartlegging av hvordan FNs bærekraftmål implementeres i regional og kommunal planlegging* [Mapping of SDG implementation in regional and municipal planning] (NF-report 7/2020). Nordlandsforskning.

Miljøverndepartementet. (1988). *Miljø og utvikling. Norges oppfølging av Verdenskommisjonens rapport* [Environment and development. Norway's follow-up of the World Commission's report] (Meld. St. nr. 46 (1988–89)).

Miljøverndepartementet. (1991). *Om miljøvern i kommunene* [On environmental policy in the municipalities] (Meld. St. nr. 34 (1991–92)).

Miljøverndepartementet. (1993). *Tenke globalt—handle lokalt. Nasjonalt prioriterte satsningsområder for det kommunale miljøvernarbeidet* [Think globally—act locally. Nationally prioritized areas of focus for municipal environmental protection work] (T-937).

Myrhaug, E. (Ed.). (1987). *Økokommunen—tenke globalt, handle lokalt* [The eco-municipality—think globally, act locally]. Venstres Opplysningsforbund.

Næss, P. (1992). *Natur- og miljøvennlig tettstedsutvikli*ng [Nature- and environment-friendly urban development] (NIBR Rapport 1992:2). NIBR.

Naustdalslid, J., & Hovik, S. (1994). Lokalt Miljøvern [Local Environmental Protection]. Tano.

Nøttestad, Ø. (1999). *Miljøvernforvaltningen i tidsperspektiv. Del III: Mot vår tid (1982-1992)* [Environmental management in perspective: Part III: Towards the present era (1982–1992)]. Ministry of the Environment.

Ramsdal, H. (1997). *Evaluering av økokommuneprogrammet* [Evaluation of the eco-municipality program] (VF-report 3/97). Western Norway Research Institute.

Vindegg, M., Christensen, I., Aall, C., Arnslett, A., Tønnesen, A., Klemetsen, M., Temesgen, A. M., Hovelsrud, G. K., & Selseng, T. (2022). *Barrierer for klimatilpasning på lokalt og regionalt nivå* [Barriers to climate adaptation at local and regional levels] (M-2263). CICERO/ Miljødirektoratet.

Chapter 3
Up Like a Lion and Down Like a Lamb—The Legacy of Minister of the Environment Torbjørn Berntsen in the Debate on Sustainable Consumption

Abstract This chapter explores the "golden age and decline" of the debate on sustainable consumption in Norway, with former industrial worker, Labour Party politician, and Minister of the Environment Torbjørn Berntsen as a central figure. The author analyzes how, in the 1990s, Berntsen challenged both the international community and Norwegian politics by questioning the necessity of economic growth and promoting the need to reduce consumption in affluent parts of the world. Drawing on a review of radical municipal pilot projects and political initiatives within the UN, the chapter highlights the dynamics between local experimentation and national paralysis. It documents how a burning political issue of global justice and resource distribution was gradually neutralized and transformed into technical questions of efficiency and reflects on the lessons from a period when political courage confronted the iron law of growth economics.

Introduction

The chapter on consumption revolves around my favorite environmental politician in Norway, former Minister of the Environment Torbjørn Berntsen, and the initiative he launched in the 1990s—first within the UN Commission on Sustainable Development (CSD) and later in Norwegian environmental policy—to bring the issue of sustainable consumption onto the political agenda. In particular, the question is whether those of us in affluent parts of the world must reduce our consumption to allow for an increase in consumption in poorer regions.

In this chapter, I will review the "rise and fall" of this debate in Norway and demonstrate the power of achieving collaboration between radical elements in local and central administration. The strength lies in using municipalities as arenas for the development of new radical political ideas and practices, while the challenge—as I will also show—is establishing a connection back to the national level to effect changes in national policy. Or, as my good friend and colleague William M. Lafferty stated in connection with the evaluation of the Norwegian local experiments to

C. Aall, *Sustainable Development*, https://doi.org/10.1007/978-3-032-35930-8_3

develop a policy for sustainable consumption: "Municipal policy is good; municipal plus state-level policy is significantly better."

The "Shitbag" incident: How Torbjørn Berntsen expanded the English vocabulary

Before entering full-time politics—First as a deputy representative to the Storting in 1965—Torbjørn Berntsen worked for 15 years as a ship pipefitter. He remains one of the few prominent labour party figures in recent decades with a genuine background as an industrial worker. This heritage defined his political persona; Berntsen was, and is, known as a man who "speaks his mind" without the filter of traditional diplomacy.

In 1990, Berntsen was appointed Minister of the Environment. He threw himself straight into the fight against "acid rain"—a major environmental crisis at the time. The pollution came mostly from sulfur dioxide from British coal combustion. Prevailing winds carried it across the North Sea, acidifying Norwegian watercourses and wiping out salmon and trout populations on a massive scale.

The tension between Norway and the United Kingdom reached a boiling point in August 1993. At a public meeting, Berntsen famously called the British Environment Secretary—later Lord—John Gummer a "drittsekk." The outburst had roots in a meeting earlier that year. When Berntsen raised the issue of British emissions, Gummer dismissively countered by attacking Norwegian whaling instead. According to Berntsen, Gummer's demeanor was not only uncooperative but patronizing toward the Norwegian government.

The incident gained immediate international notoriety when British newspapers translated Berntsen's characterization as: "John gummer is the biggest shitbag I have ever met."

While "drittsekk" (literally "dirt sack") is a common, albeit coarse, Norwegian derogatory term for someone you find unpleasant or contemptible, the English translation "shitbag" packed an even heavier punch. Following Berntsen's outburst, the term—And the story behind it—Became a legendary footnote in environmental diplomacy, illustrating the clash between the blunt language of a Norwegian pipefitter and the polished rhetoric of the British establishment.

Environmental Policy Background

The critical debate on private consumption in an environmental context can be traced back to the 1960s critique of economic growth as a goal for societal development, and how the pursuit of growth drives both increasing environmental problems and widening disparities between what were then referred to as developing countries and industrialized nations. In Norway, this debate was enriched by the emergence of a distinct philosophical movement—deep ecology and eco-philosophy—with the Norwegian professor Arne Næss as its leading proponent.[1] Advocates of deep ecology argue that all life has intrinsic value, not merely value based on its utility to humans. Therefore, humanity must adapt to the needs and demands of nature.

Equally central to the Norwegian debate was the book—and later the eponymous movement—*The Future in Our Hands* (FIVH), written by Erik Dammann in 1972. A recurring theme in his books and for FIVH is that the future is shaped by the actions of individuals and society, and that it is both necessary and possible for people in the industrialized world to adopt a new lifestyle with significantly reduced consumption.

As discussed in the chapter on the concept of sustainability, the 1987 report of the World Commission does not pursue the critique of economic growth that was so central to the socially critical segment of the environmental debate, both in Norway and internationally. Nor does the report address private consumption as a driver of unsustainable development. Instead, attention is directed toward environmental problems associated with the production of goods and services.[2] The UN's action plan for sustainable development from 1992—Agenda 21—*does* address consumption. Chapter 4, titled "Changing Consumption Patterns," outlines rather general objectives regarding consumption in industrialized countries (§ 4.2, emphasis added):

> Poverty and environmental degradation are closely linked. While poverty contributes to certain forms of environmental stress, the main cause of the ongoing deterioration of the global environment is the unsustainable *pattern* of consumption and production, particularly in industrialized countries. This *pattern* is a cause for serious concern. It exacerbates poverty and increases imbalance.

It is evident from the context that the text problematizes final consumption by consumers (individuals and households), rather than society's total resource use. Nevertheless, this represents a new perspective in official environmental policy and aligns more closely with viewpoints advanced by a critical segment of the environmental movement. The key question, however, is how we should interpret the term "pattern." Does this refer to the volume of consumption or merely its composition? And whose consumption, specifically, must change?

As early as 1993, one year after the Rio Conference, the national assembly of the Norwegian Association of Local and Regional Authorities (KS) adopted a statement that offers an even more radical approach to understanding sustainable consumption (p. 9, my emphasis)[3]:

> Our *high consumption* and *economic growth in the West* are incompatible with sustainable development and the equitable global distribution of the world's resources. In this context, our consumption patterns are a key issue. Municipalities and county authorities must take the lead in actively promoting consumption patterns that are in line with natural resources and ecological limits.

---

[1] Næss (1976).

[2] Høyer (1997).

[3] Kommunenes Sentralforbund (1993).

Like Agenda 21, the KS also employs the term "consumption patterns." However, the radical element in the resolution from KS is its critique of the goal of economic growth. The natural interpretation is that changing consumption patterns entails reducing (or limiting) consumption in economic terms. This perspective does not necessarily disregard the potential to stretch resources through energy efficiency, recycling, product downsizing, substitution, and so forth. Rather, it suggests that these measures alone are not assumed to be sufficient—in the long run and in a sustainable manner—to provide all the world's people with a level of consumption comparable to that of present-day Norway.[4]

At the first meeting of the UN CSD following the Rio Conference in 1992, then Minister of the Environment Torbjørn Berntsen proposed placing sustainable consumption on the agenda for the subsequent UN CSD meeting. At the same time, he stated that Norway could host a preparatory meeting on this topic and, through national pilot projects, demonstrate what efforts to promote sustainable consumption in high-income countries like Norway could concretely entail. The proposal was accepted, and thus the first international meeting on sustainable consumption with participation at a high political level was held in Oslo on 19–20 January 1994. Over time, Norway would host several international meetings on this topic (see Footnote 4).

In February 1993, the Norwegian Labour Party government presented the long-term program for the 1994–1997 period, which projected, among other things, a doubling of per capita consumption by the year 2030. One year after the presentation of that program, during the opening of the Oslo-conference, Torbjørn Berntsen made the following statement, which quickly attracted significant media attention, both nationally and internationally (*Aftenposten*, 19 January 1994):

> One thing is certain: The goal of doubling Norwegian consumption by 2030 is completely unrealistic. If India were to pursue the same ambitions as we do, the world would collapse. We Norwegians must recognize that there is no room for further growth in prosperity. The problem is that no matter how environmentally friendly Norwegian industry becomes, and no matter how sustainable our products are, any increase in growth in the North contributes negatively to global equity.

The radical aspect of Berntsen's statement is his emphasis on the necessity of reducing the volume of consumption, and on how resource efficiency and changes in the composition of consumption are insufficient to achieve sustainable consumption. In the debate on the long-term program in the Storting on 3 March 1994—three months after Berntsen made his (in some eyes) infamous remarks at the opening of the international conference on sustainable consumption—all the "small" parties in the Storting (the Centre Party, Socialist Left Party, Red Electoral Alliance, Christian Democratic Party, and Liberal Party) jointly proposed an additional report. In that report, the government would present a policy and strategy for reducing Norway's consumption of natural resources, directly referencing the interview with Berntsen in *Aftenposten*. In the subsequent debate, Berntsen's perspective was clearly rejected

[4] Aall et al. (1999).

by the Conservative and Progress parties.[5] The then leader of the Labour Party, Thorbjørn Jagland, stated that he shared Berntsen's viewpoint on the one hand, but on the other emphasized: "It is... not enough to look at how much consumption increases to assess whether it is sustainable; we must look at what kind of consumption is increasing." The chair of the Finance Committee from the Labour Party, Karl-Erik Schjøtt-Pedersen, went even further in moderating and downplaying Berntsen's statements, arguing that "it will not be possible to maintain the welfare systems we have in society without ensuring economic growth." Schjøtt-Pedersen later had the opportunity to follow up on this political stance in his subsequent role as director of the industry organization Norwegian Oil and Gas from 2015 to 2020. The proposal from the five opposition parties was defeated by the votes of the Conservative, Labour, and Progress parties.

Round One in the Battle for Sustainable Consumption: Berntsen Lost—Gro Won

A key metric for private consumption is the "consumption in households and non-profit institutions" statistic compiled by Statistics Norway (Table 09842). In 1993, per capita consumption in Norway stood at NOK 98,661. By 2024, that figure had risen to NOK 368,018. These values are recorded in current prices. To assess the actual change in volume over time, you must adjust the figures to constant prices. When you do that—adjusting for inflation and projecting linearly to 2030 based on the 1993–2024 growth rate—the data indicate a 95% increase. This aligns remarkably closely with the objective of doubling consumption between 1993 and 2030, as established in the government's long-term program adopted by the Storting in 1994.

So, the then minister of the environment, Torbjørn Berntsen from the labour party, lost the opening round in the struggle for sustainable consumption in Norway. Meanwhile, his former colleague and fellow party member Gro Harlem Brundtland—Former minister of the environment and then prime minister—Emerged victorious, securing the path for continued economic growth.

As a follow-up to Berntsen's commitment to the UN CSD, several initiatives were launched in Norway in the mid-1990s to help clarify and define the concept of "sustainable consumption." This is where I became involved, as project manager for the first project at Western Norway Research Institute (WNRI) focused on sustainable consumption—an evaluation of one of the initiatives started by Berntsen. Over time, we formed a small core group of researchers in this field, including Karl Georg Høyer, Erling Holden (who completed his doctoral research on this topic at WNRI), and not least John Hille (originally affiliated with the Idébanken Foundation, but who, in the following years, held a number of project positions at WNRI).

---

[5] Stortinget (1994, 3. mars).

A Study of Barriers to the Transition to Sustainable Consumption

During the 1995–1999 period, seven municipalities—Steigen, Røros, Flora, Stavanger, Kristiansand, Hurum, and Fredrikstad—participated in the Sustainable Local Communities project, funded by the former Norwegian Pollution Control Authority and established as a direct consequence of Berntsen's commitment to the UN CSD. The project concluded with an evaluation, which I led. In the evaluation, we aimed to determine how the municipalities understand and choose to approach the task of achieving sustainable consumption, what obstacles they encounter in their work on sustainable consumption, and, finally, to propose ways to overcome the identified barriers. The municipalities implemented nearly 100 sub-projects over four years, and we conducted close to 60 interviews with involved stakeholders. Thus, we had access to a rich body of empirical material.

The *first* research question concerned the municipalities' *understanding* of sustainable consumption. The analysis was based on a four-level "scale" ranging from "no" to "great" emphasis on the understanding expressed by Berntsen in his widely discussed interview with *Aftenposten*:

- Category 1: No emphasis on consumption: environmental policy focuses exclusively on reducing the environmental impact from the production of goods and services for private consumption.
- Category 2: Improving consumption efficiency: for example, encouraging the use of private cars that consume less fuel per kilometer..
- Category 3: Changing the composition of consumption: for example, promoting a shift from private cars to public transportation.
- Category 4: Reducing the overall level of consumption: for example, land-use planning aimed at decreasing the extent of transport through urban densification and reducing the distance between workplaces, social service centers, commercial hubs, and residences.

On the one hand, we found that the subprojects initiated by the municipalities were, to a small or negligible extent, purely production-oriented (category 1), and that the efficiency perspective was also scarcely represented (category 2). On the other hand, few measures explicitly aimed to reduce private consumption in specific areas (category 4). Most of the measures focused on changing the composition of consumption (category 3), but there were also examples of initiatives that sought to address the overall scale of consumption, particularly related to transport volume and total energy use in households.

The project managers in the municipalities acknowledged at the conclusion of the project that the work had largely focused on positive and uncontroversial topics to achieve tangible results within the project's short timeframe. At the same time, a clear awareness existed that the project's emphasis on sustainable consumption points toward the significant and controversial environmental challenges associated with reducing consumption levels in affluent parts of the world. Several informants highlighted the necessity of achieving what they described as a "quantum leap" in environmental policy, and they expressed hope that the time would soon be ripe to address these much more controversial aspects of efforts toward sustainable development at the local level (see Footnote 4).

The Røros Paradox: When "Deep Green" Tourism Met the Distance Barrier

Røros is one of Europe's oldest wooden towns and among the few mining communities worldwide inscribed on the UNESCO world heritage list. As part of the sustainable local communities project, the municipality aspired to transform Røros into Norway's premier sustainable tourism destination. This ambition included addressing the environmental footprint of travel to and from the region.

Initially, the focus was on strengthening the environmental profile of local activities. This meant transitioning to locally sourced organic ingredients, expanding public transport for visitors, and promoting a traditional, non-motorized Norwegian winter vehicle: The *sparkstøtting* (kicksled).

Once these local initiatives were up and running, the municipality tackled the bigger systemic challenges. Then something unexpected happened. The airline Braathens S.A.F.E.—Which served the small local airport—Agreed to a municipal request to consider reducing their flight schedule. Conversely, and perhaps equally surprisingly, the state railway company (*Norges Statsbaner* [NSB]) declined a request to increase train frequency.

The final hurdle, however, came from abroad. When the municipality approached a German travel agency specializing in "deep green" tourism, they received a polite but firm rejection. The agency's response highlighted a fundamental dilemma: While the local product was exemplary and the environmental profile impeccable, Røros was simply too far from their customer base. The agency's core principle was promoting short-distance holidays—And for a German eco-tourist, a trip to the Norwegian mountains was a bridge too far.

The *second* research question concerned the barriers municipalities encounter in their efforts toward sustainable consumption. Analyzing barriers related to local development initiatives was relatively new within environmental management; therefore, we first needed to develop a functional methodology. We based our approach on three theoretical frameworks: James D. Thompson's distinction between clear and unclear goals and means[6]; Paul Sabatier's theory on barriers in public administration and policy implementation[7]; and Evert Vedung's analysis of rationality in

[6] Thompson (1967).

[7] Sabatier (1986).

the use of goals and instruments.[8] Based on these, we developed a framework for the analysis of barriers, which has proven useful in several subsequent projects for environmental management.[9] The typology distinguishes between the dimensions of "degree of conflict" and "level of barriers." For the conflict dimension, we differentiate between ambiguous and clear policy instruments, as well as ambiguous and clear objectives. The term "ambiguous" refers to "disagreement," while "clear" indicates local consensus regarding which policy instruments to use or which objectives should underpin the choice of instruments. This provided a basis for identifying three degrees of conflict:

- Trivial obstacles: situations where there is agreement on both means and objectives.
- Unclear means: situations where there is agreement on objectives but disagreement regarding the means to achieve them.
- Goal conflicts: situations where there is disagreement about objectives.

In addition, we distinguished between five levels in the governance hierarchy where the barriers may manifest:

- Individual level: the individual resident.
- Local level: organizations and businesses within the municipality.
- Municipal level: within the municipal organization.
- National level: within the state.
- Global level: supranational governing bodies.

The combination of goal conflicts and the national or global level represents the type of barrier that a municipality will typically find the most challenging to circumvent or mitigate. It is therefore perceived as the most significant form of obstacle. Municipal barriers are generally much more within a municipality's capacity to influence, while local and individual barriers fall under an intermediate category.

Table 3.1 presents examples of specific barriers identified by our informants and how these are distributed across the various categories of obstacles. The table demonstrates a clear correlation between the degree of conflict and the level within the governance hierarchy to which the conflicts can be attributed: the higher up in the hierarchy, the more contentious these are perceived by the municipalities. In approximately 70% of cases where municipalities experience the barriers as serious goal conflicts, these are linked to the national or international level. Conversely, municipalities report that 85% of the "trivial" barriers are individual or local.

Municipal barriers primarily (80%) concern ambiguities related to policy instruments. Trivial barriers are mainly associated with the local community (the local and individual level). Lack of local support and engagement from households, as well as insufficient dialogue between local communities and the municipal organization, appear as recurring issues. Unclear policy instruments were largely related to internal

[8] Vedung (1998).

[9] Vindegg et al. (2022).

Table 3.1 Distribution of and examples of obstacles experienced by 7 municipalities in 60 sub-projects on sustainable consumption (see Footnote 4)

Barrier level	Degree of conflict		
	Trivial barriers	Unclear instruments	Goal conflicts
Individual barriers	20% e.g., lack of knowledge about alternative energy systems	5% e.g., weak recruitment to the fishing profession	3% e.g., difficult to change attitudes among youth regarding clothing purchases and driving
Local obstacles	16% e.g., lack of media interest in environmental protection	4% e.g., disagreement on the importance of pursuing environmental certification.	4% e.g., opposition to removing private parking spaces
Municipal barriers	4% e.g., lack of knowledge about alternative energy	20% e.g., absence of municipal energy and climate policy	1% e.g., conflicts between farmland protection and residential development near city centres
National barriers	3% e.g., lack of information about subsidies for alternative energy	1% e.g., food safety authority requirements for the food industry	12% e.g., low prices for electricity and oil
Global barriers	> 0% e.g., EEA regulations prevent prioritization of local suppliers	3% e.g., EEA requirements for hygiene in slaughterhouses for export	4% e.g., EEA prevents local marketing of fish products

governance challenges within the municipal organization, while goal conflicts were predominantly associated with the national level.

We also found a correlation between the type of barrier and municipality type: smaller rural municipalities experienced a higher degree of conflict than larger urban municipalities.

The *third* research question concerned what can be done to circumvent or reduce the identified barriers. Evert Vedung distinguishes between pilot activities as *social experiments* and as *implementation strategies* (see Footnote 8). In the latter case, pilot activities are essentially used to achieve specific objectives. Given Berntsen's underlying goal with the "Sustainable Local Communities" project, it was natural to adopt the implementation strategy perspective.

Our main conclusion was that the project generated many relevant experiences indicating a need for adjustments in national policy across several areas. However, feedback from these experiences relevant to national policy occurred to a limited or negligible extent. Production-oriented environmental policy continued to dominate throughout the 1990s and well into the 2000s, with few initiatives aimed at changing consumption patterns.

Oslo as the First City in the World to Undertake a Scientific Analysis of Its Residents' Ecological Footprint

A significant new step in the Norwegian history of sustainable consumption was taken in 2001 when the City of Oslo commissioned an analysis of the local ecological footprint (EF). This initiative was part of the effort to revise the municipality's Urban Ecology Programme for the 1990–2010 period. That program remains the first—and so far only—municipal planning initiative of its kind in Norway. Until then, the urban ecology program focused primarily on distinctly local environmental issues, such as air quality in the city and water quality in local watercourses and the Oslofjord. By employing the EF tool, the municipality aimed to shed light on more pronounced global environmental challenges, particularly those related to the scale and composition of private consumption in Oslo.

Oslo: Europe's Greenest Capital in Both Senses of the Word

The European Environment Agency (EEA) ranks Oslo as the greenest capital in Europe in physical—or more precisely, "biological"—terms. Measured by tree cover, Oslo boasts an impressive 72%, vastly outperforming the 30% average across Europe's 38 capital cities.

However, Oslo's "greenness" is equally prominent in a political sense. As early as 2003, the city received the European Sustainable City award in recognition of its pioneering urban ecological program.

Sixteen years later, in 2019, Oslo was officially named the European green capital. This prestigious title followed a significant political shift in 2015, when the Green Party (Miljøpartiet De Grønne) joined a governing coalition in a Norwegian municipality for the first time (a milestone shared with Trondheim that same year). The inaugural declaration of Oslo's new "red-green" city council—Comprising the Labour, Socialist Left, and Green parties—placed a car-free city center at the top of its political agenda.

The project emerged through a collaboration between the City of Oslo, the Programme for Research and Documentation for a Sustainable Society (ProSus), and WNRI. In the foreword to the report, the head of ProSus, Professor William M. Lafferty, situates the project within the ongoing Norwegian and international debate on sustainable consumption (p. 3)[10]:

> Although the debate surrounding the issue of sustainable production and consumption has subsided significantly in Norway, it remains a fundamental aspect of the UN's work on sustainable development and will receive considerable attention at the World Summit on Sustainable Development ("Rio+10") in Johannesburg, August 2002. The purpose of the report presented here is to ensure that the issue of sustainable production and consumption also receives greater attention in Norway's capital, Oslo.

---

[10] Aall and Norland (2002).

The concept of the "ecological footprint" (EF) was developed by William Rees and Mathis Wackernagel at the University of British Columbia in the 1990s.[11] The fundamental idea behind the EF is that while the Earth's land area is limited, the human population is increasing, along with land use. All human activities—and all natural resource consumption—require land area, primarily biologically productive land. Unlike analytical tools that were common at the time, such as life cycle assessments, the EF also incorporates the concept of global equity and the aspect of the "volume" of consumption (see Footnote 10). The latter entails increased attention to consumption and its associated environmental impacts. Local consumption is put into perspective by highlighting the planet's available natural resources and assimilative capacity.

This was not only the first time the EF method was applied in a Norwegian study but also one of the first instances internationally where the method was transferred from the national to the municipal level.[12] Previously, it had only been used in more limited local analyses in Finland,[13] the Netherlands,[14] and Austria.[15]

The report provided a range of new insights that contributed to establishing a knowledge base for developing what would become one of the most ambitious environmental programs for a European capital. On the one hand, the analysis confirmed the significance of the three major Bs— Biff, Bil, Bolig (beef, car, and housing)—in typical high-consumption countries. Food accounted for the largest share (50%), while transport- and housing-related consumption each represented about 25%. Air travel alone accounted for 14%.

On the other hand, we found it more surprising that the total EF of Oslo's residents was only 1% lower than the national average for Norway. This means that Oslo's environmental advantage, assumed by many to stem from smaller living space per person and a lower car ownership rate (and thus less private car use) compared to the national average—is offset by higher consumption of air travel. The significantly lower use of private cars for daily travel was, to some extent, counterbalanced by higher consumption related to leisure travel than the national average. The EF associated with food consumption also proved to be somewhat higher than the national average in Norway.

The areas of action derived from our analysis differed from what was commonly found in municipal environmental plans at the time in several key respects: food consumption, leisure travel by private car and airplane, and the size of residential living space.

In local environmental policy contexts, the "food" area of action was typically associated with the environmental impact of food production and was therefore considered less relevant for urban municipalities. We instead highlighted local food

[11] Wackernagel and Rees (1996).

[12] Aall and Norland (2005).

[13] Hakanen (1999).

[14] De la Court et al. (2000).

[15] Daxbeck et al. (2001).

consumption, with measures aimed at shifting local diets toward more plant-based foods and placing greater emphasis on organic and locally sourced products.

Leisure travel by private car often takes place outside the municipality of residence, with trips to holiday cabins being predominant. As a result, this type of activity frequently falls outside the scope of local environmental policy. The exception is municipalities that are destinations for such leisure travel—for example, typical "cabin municipalities". However, we emphasized that the City of Oslo should develop measures to reduce the volume of leisure travel out of the municipality, or alternatively, to shift travel from private cars to public transport. The same recommendation applied to leisure travel by air; however, in this area, there are limits to what a municipality can achieve on its own. This is especially true for Oslo, from which the country's main airport is only 20 min away by express train.

An area over which cities have significantly greater influence is urban form, which can be managed largely through land-use planning. EF calculations highlight an important relationship: living space per person, in addition to being an indicator of high resource consumption, determines housing-related consumption. This latter applies not only to household operations (as a large floor area requires more energy for heating) but also to the available space for accumulating consumer goods (such as furniture and electrical appliances).

In our study, we pointed out that the consumption of *living space* in Norway has increased by 43% over the past 20 years, and that the living space per capital was roughly 25% higher in Norway than in comparable industrialized countries. Our analysis identified the principle of *decentralized concentration* as an environmentally sound model for further urban development in Oslo. This involves establishing compact mini-cities within Greater Oslo. Subsequent studies of the City of Oslo[16] as well as Helsinki[17] have confirmed that such strategic measures—when implemented—have a positive environmental effect.

The Brundtland Report Celebrates 20 Years

In the winter of 2007, the idea emerged to prepare an anniversary report in connection with the 20th anniversary of the Brundtland Report. The aim was to revitalize the debate that Berntsen had initiated over a decade earlier regarding the necessity of reducing consumption in wealthy countries. The idea also stemmed from a complete rejection of consumption-oriented measures by the government-appointed Low Emissions Commission, established in 2005 with the mandate to assess how Norway could reduce its national greenhouse gas emissions by 50–80% towards 2050. The commission was chaired by Professor Jørgen Randers, one of the authors of the seminal report *Limits to Growth.*

---

[16] Holden and Norland (2005).

[17] Tiitu et al. (2019).

As part of the committee's work, the Idébanken Foundation and John Hille were commissioned to assess climate measures aimed at changing Norwegian consumption patterns. The assessment was referenced in the committee's report but was not published alongside other supporting documents on its website (which have since been removed).

The committee submitted its report in 2006[18] and concluded that it was "necessary, feasible, and not prohibitively expensive to reduce Norway's greenhouse gas emissions by two-thirds by 2050" (p. 11). In the foreword, Randers wrote the following regarding an omission from the report (p. 11):

> A radical transformation of Norwegian lifestyles towards a more climate-friendly direction could significantly reduce future emissions. Nevertheless, the committee has chosen not to recommend this, partly because we believe it would be an impossible political task to implement.

The idea for a commemorative report marking the 20th anniversary of the Brundtland Report was initiated by Jacob Bomann-Larsen, then serving as Special Advisor at the Ministry of Children and Equality. He was previously a co-founder, together with Erik Dammann, of the organization FIVH, which since its inception in 1974 has been the most prominent voice in the Norwegian debate on sustainable consumption. Those of us who received and further developed the idea into a project included John Hille at the Idébanken Foundation, and Karl Georg Høyer and myself at WNRI. This led to WNRI being entrusted with the first government-funded study of how Norwegian private consumption impacts the environment—two years after the government's Low Emissions Commission had so clearly rejected such an idea. The assignment was commissioned jointly by two ministries: the Ministry of Children and Equality, based on its responsibility for consumer affairs, and the Ministry of the Environment.

The analysis used two academic reference points. More than ten years earlier—in 1995—John Hille had been commissioned by the Alternative Future Inquiry to analyze the development of Norwegian consumption in relation to global resource and mineral availability into the next century. He was also asked to address the question of what level of consumption Norway could maintain if developing countries were to achieve the same standard of living as Norway by the year 2025.[19] Ten years later, the Swedish Ministry of Finance published an official report on consumption-oriented Swedish environmental and climate policy[20] using the same alliteration—beef, car, and housing—that we had introduced a few years earlier.[21]

Our analysis yielded three main conclusions regarding developments in the 20 years since the Brundtland Report was published[22]:

- The environmental impact of Norwegian production has decreased somewhat.
- The environmental impact triggered by Norwegian consumption has increased.

---

[18] NOU 2006:18 (2006).

[19] Hille (1995).

[20] SOU 2005:51 (2005).

[21] Aall (2002).

[22] Hille et al. (2008).

- The environmental impact of transport has increased significantly.

We had expected to find a greater reduction in the environmental impact from Norwegian *production*. We identified two reasons why the development had not been as positive as anticipated. The *first* reason concerns a uniquely Norwegian circumstance compared to other OECD countries: namely, that Norway, at the time our analysis was conducted (and, as it turned out, continuing thereafter), still had a large share of energy-intensive industrial production. This, in turn, means that the environmental impact of Norwegian industrial production has not declined as sharply as in other OECD countries. The *second* reason relates to greenhouse gas emissions from Norwegian international shipping. Our analysis revealed that the official Norwegian emission inventories significantly underestimated these emissions. Based on our calculations on the proportion of Norwegian-owned vessels in the total international tonnage, we estimated that greenhouse gas emissions from Norwegian international shipping could be four to six times higher than the official figures at the time. This finding necessarily alters the overall picture of the environmental impact of Norwegian production accordingly.

The analysis further indicated that changes in leisure consumption particularly contributed to an increase in the environmental impact of Norwegians' total private consumption. This conclusion drew on a previous national survey that specifically mapped Norwegians' leisure consumption.[23] This was the first time in Norway—and, as it turned out, internationally—that a systematic survey had been conducted on the scope and composition of a country's leisure consumption, measured in turnover, physical quantities, and total direct and indirect energy use.[24] The analysis yielded some surprises and confirmed certain assumptions (see Fig. 3.1). The confirmation concerned the extent of energy use triggered by Norwegians' air travel (see the share of mobile energy use for "holiday journeys" and "visiting friends and relatives"). The surprise related to the extent of energy use for "electronic home entertainment" and "outdoor recreation."

The latter finding led to some skepticism from representatives of outdoor recreation interests. It was evidently difficult to acknowledge that the proud and, by definition, environmentally friendly Norwegian outdoor recreation tradition could constitute part of the environmental problem and no longer be synonymous with the environmental solution. This observation gave rise to the term "the outdoor recreation paradox." The substantial energy use associated with outdoor recreation was distributed roughly equally between travel to and from the activity sites and the consumption of equipment, which has come to play an increasingly prominent role in the practice of outdoor recreation. One of my master's students updated the figures from 2007 using the same methodology.[25] This more recent study suggests that the energy use associated with outdoor recreation has remained roughly constant in the period from 2001 to 2017, while energy use for leisure air travel has nearly doubled.

[23] Hille et al. (2007).

[24] Aall (2011).

[25] Rønnevik (2019).

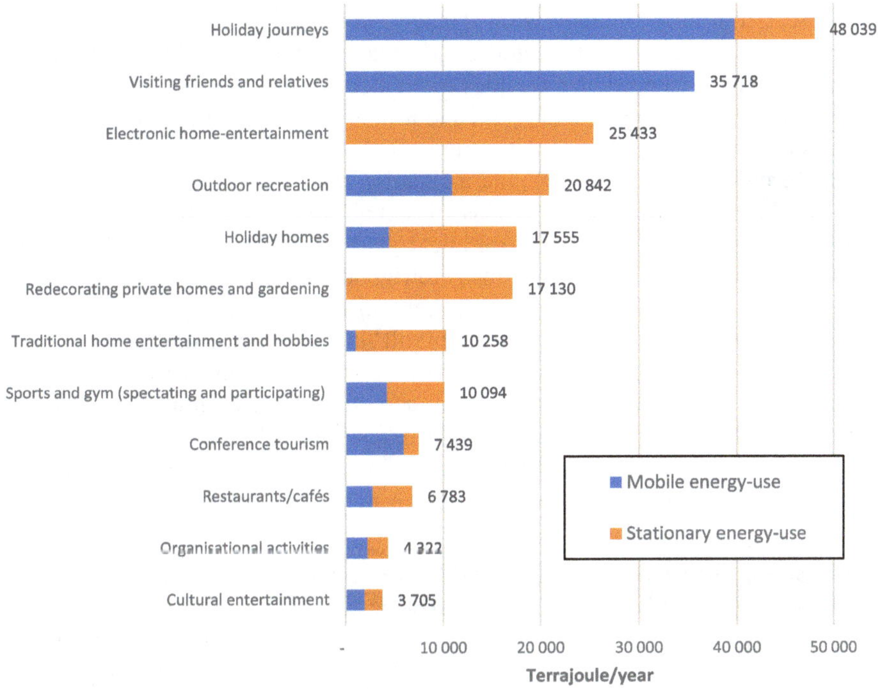

Fig. 3.1 Total direct and indirect energy use triggered by different categories of Norwegians' leisure consumption in 2001, divided into mobile and stationary energy use. See Hille et al. (2007)

Although consumption related to outdoor recreation had increased significantly in the 20 years since the Brundtland Report was published (and likely has continued to rise in some areas since then), and although certain forms of leisure consumption are more energy-intensive than everyday consumption, leisure consumption accounts for only 25% of the total direct and indirect energy use triggered by Norwegians' private consumption. This observation inspired me to formulate a hypothesis that leisure consumption could become part of the sustainability solution.[26] This hypothesis stood in stark contrast to the prevailing view of leisure consumption as "conspicuous consumption" serving the social function of displaying a person's wealth, status, or social position—a concept established more than 100 years ago by the American economist and sociologist Thorstein Veblen in his book *The Theory of the Leisure Class* (1899).

My competing hypothesis was based on the assumption that as consumers, we are particularly open to new ideas and to learning new practices during our leisure time—and perhaps especially when we are on holiday far from home in new social settings. Thus, even if the leisure activity itself (for example, a flight to a Greek holiday island) entails a significant environmental impact, this impact could be more

[26] Aall et al. (2011).

than offset if, during the holiday, we are "taught" or otherwise inspired to bring home more environmentally friendly attitudes and behaviors, which are then implemented in our everyday consumption. Examples could include adopting new and more environmentally friendly dietary habits (such as a vegetarian diet) or modes of mobility (such as increased use of bicycles in daily life).

On the other hand, let us imagine a situation where public consumption—for various political and ideological reasons—is not allowed to grow further, and where everyday consumption (or at least the physical goods we purchase) has reached a saturation point. (There is a limit to how many refrigerators, TVs, furniture items, and so on we find it meaningful to own.) In such a scenario, leisure consumption becomes the last category of private consumption that can still "lubricate" the economic machinery to continue delivering economic growth. If this is the case, any serious attempt to reduce the environmental impact of leisure activities that also limits the economic volume of such consumption will face massive resistance. This resistance will come not only from the businesses and individuals directly affected but also from the political and economic power holders who depend on continued growth.

The Second Attempt to Put Sustainable Consumption on the Agenda: How Anniken, Erik, and Karl Georg lost—And Jens Won

On 23 March 2008, John Hille and I arrived at a meeting room in the Ministry of Children and Equality. Our mission was to present our anniversary report to the two ministers that formally commissioned the report: Anniken Huitfeldt (Minister of Children and Equality, Labour Party) and Erik Solheim (Minister of the Environment and International Development, Socialist Left Party). In a joint press release, the ministers expressed a rare level of ambition regarding the consumer's role:

> I wanted to know what specific changes are needed in our daily lives so that we, as consumers, can reduce the environmental and climate impact of consumption in Norway. Environmental and climate concerns must have consequences for consumption. Previously, measures primarily targeted production; now, it is vital to also focus on consumption (Huitfeldt).
> We live in a world where national borders matter less and less. Environmental impacts cross borders just as goods and people do. This study helps demonstrate how Norwegian consumption affects the global environment (Solheim).

By the autumn of 2008, Karl Georg Høyer at WNRI was asked to chair the public committee for which our report was intended to provide the foundation. The committee's mandate was to assess orienting climate policy toward consumption, thereby supplementing the 2006 Low Emissions Commission report. Throughout that autumn, Karl Georg contributed to drafting the mandate. However, the formal appointment kept getting delayed. Then, in June 2009, Karl Georg received informal word that the cabinet—led by Prime Minister Jens Stoltenberg—had decided to scrap the committee entirely. The decision came during the final government conference before the summer break. One can only speculate whether the fear of losing the upcoming autumn election made the prospect of a "consumption-focused" climate policy too politically radioactive.

Experiences and the Way Forward

My research specifically on the topic of sustainable consumption concluded in 2008 with the anniversary report marking the publication of the World Commission's report, which we conducted on behalf of the Ministry of Children and Equality and the Ministry of the Environment. The public debate under the heading "sustainable consumption" also faded toward the end of the 2010s, or rather it became limited to the question of how consumption is addressed within climate policy. In that context, both our research and the public debate flourished, as I describe in the chapter on climate policy.

What insights have I gained after more than ten years of research—from the early 1990s into the 2000s—regarding the broader issue of sustainable consumption? Which of these insights may also be relevant to the ongoing debate on consumption-oriented climate policy? Here are my four main points:

- Politics is a distinct craft, and alliance-building among radical forces across levels of governance can change policy.
- "The Norwegian way of life is not up for negotiation."
- Technological measures are strongly preferred over changes in consumption.
- The hope for achieving more sustainable consumption lies with the municipalities—not with national politicians.

Already in the introduction to this chapter, I expressed my admiration and respect for former Minister of the Environment Torbjørn Berntsen. He demonstrates to me that *politics* is a *craft* that extends far beyond acquiring subject-matter expertise through formal education. Many ministers of the environment who succeeded Berntsen possessed varying degrees of formal or informal expertise in environmental issues that far exceeded his own (he was a plumber!). Yet in my view, he remains the Norwegian minister of the environment who came closest to truly shifting national policy in a new and more sustainable direction.

The political processes Berntsen initiated in the area of "sustainable consumption" also illustrate the significant potential for change that lies in establishing alliances across levels of governance among actors oriented toward transformation. At the same time, the experiences show that implementing pilot initiatives is no guarantee of achieving lasting change. In the chapter on local environmental protection, I describe how such a pilot scheme—the Environmental Protection in Municipalities project—paved the way for lasting change in the form of a reform bearing the same name, but which lost much of its radical transformative potential along the way.

The same mechanism is illustrated in the case of sustainable tourism, where a project on the sustainability certification of tourism destinations was followed up by the establishment of a developmental scheme. However, the content of the scheme has not been sufficiently robust to influence the development of tourism sector sustainability to the degree originally envisaged (see the chapter on sustainable tourism).

In the area addressed by this chapter—sustainable consumption—the pilot schemes *did not* lead to changes in actual policy. Norwegians' consumption continues to rise toward 2030, just as predicted in the early 1990s.

The international debate on sustainable consumption, which began with the first international conference in Oslo in 1994, has undergone a shift in terminology—from initially "sustainable consumption," through "sustainable consumption and production" or "sustainable production and consumption" (the order of the terms varied somewhat), and finally to the UN Sustainable Development Goals adopted in 2015, specifically Goal 12: "Responsible Consumption and Production". The critical question is whether these changes in terminology also reflect changes in political content.

What is clear in the development of the concept is that consumption as a standalone topic proved too politically challenging. It required a connection to "production" to become politically acceptable. A reasonable conclusion is that the emphasis on purely technological measures has become greater than it would have been if "consumption" without the link to "production" had persisted—that is, if the concept had continued in its original form as "sustainable consumption." This, in turn, means that the greatest emphasis has been placed on the first level in a hierarchy of measures regarding changes in consumption: (1) increasing efficiency, (2) changing composition, and (3) changing the scale.

The Socialist Left Party: From Critique of Growth to Technological Fascination

In 1996, a significant shift occurred in Norwegian environmental politics. Odd Einar Dørum resigned as leader of the Liberal Party, marking the end of the party's era as a radical, left-leaning environmental pioneer. I had worked closely with Dørum since 1988, both on the party's environmental committee and as editor of the *Environmental Yearbook* (*Miljøårboka*). After his resignation—and the party's subsequent drift toward the right—I moved to the Socialist Left Party (SV).

However, SV was also undergoing a metamorphosis. The party was transitioning from a tradition of eco-populism and systemic critique of capitalism and economic growth towards a "light green" profile. This shift coincided with SV joining a government coalition for the first time in 2005, alongside the Labour and Centre parties. Under the leadership of Kristin Halvorsen (Minister of Finance, 2005–2009), the radical edge of consumption critique was increasingly replaced by a fascination with technological solutions.

In 2010, I attended the annual county meeting of Sogn og Fjordane SV. From the podium, Halvorsen invited input on how to modernize the party's environmental policy. To illustrate her vision, she shared an anecdote from the 1995 election campaign. Back then, the party had acquired one of the few mobile phone models available at the time—a bulky device that consumed considerable energy. Holding up her sleek, modern smartphone, she declared: "Now, we have small phones that use only a fraction of the energy. This illustrates the kind of environmental policy I want."

I took the floor and pointed out the flaw in her logic: The total energy required to produce and operate the millions of energy-efficient phones in 2010 was many times greater than the total energy used by the few "bulky" phones of 1995. This is the classic rebound effect in practice. Halvorsen looked at me, seemingly uncomprehending. Then she muttered: "That's not what I had in mind."

So, what can be learned from efforts to develop policies that specifically address the necessity of not only making consumption more efficient but also reducing it in high-income countries such as Norway? My experience suggests that resistance is too strong at the national level—even in Norway—both from within national politics and from political pressure exerted by economic actors who view such policies as undesirable.

After Berntsen's tenure as Minister of the Environment, the Labour Party became influenced by neoliberal trends. Norway also experienced several periods with various coalitions of conservative governments—including administrations with ministers from the only openly climate-skeptical political party in Norway, the Progress Party.

For Norway to adopt policies that truly reduce consumption—rather than merely shifting the composition of particularly environmentally harmful forms of consumption—I believe we need to create political pressure from below in a sufficient number of municipalities. This was what Berntsen and others sought to achieve, but so far it has not succeeded.

Berntsen Still Speaks his Mind: Twenty Years After Leaving National Politics

1997 was the last year Torbjørn Berntsen served in the Storting or as a member of the Norwegian government. However, he has continued to make his mark on public debate at regular intervals, often on environmental and sustainability issues. His most recent significant media appearance was an interview with *Magasinet for fagorganiserte* (Magazine for Union Members) in 2019.[27]

He opened the interview by praising Greta Thunberg's commitment. Then he reflected on the historical arc of the labour movement:

> I was watching the TV series about Einar Gerhardsen [former prime minister from the labour party and active politician from the 1920s well into the 1960s] the other evening. There were campaign posters from the interwar period showing huge factory chimneys spewing out terrible smoke. I thought about Thunberg's statement—*How dare you*—and thought: Einar, how could you do something like that? But we must not forget the historical process. What would our homes have been like without industrialization? We didn't know much about pollution back then. That awareness only arrived in the 1980s, with the Brundtland commission.

He then delivered one of his characteristically blunt remarks—the kind that earned him the nickname "the mouth"—criticizing the persistent obsession with economic growth:

> The whole setup is about getting more in your pocket. Buy more, use, throw away, faster and faster. Does it make any sense? When the breadbox was empty, it made sense. But is there any sense in it when you can't even get the lid back on the breadbox?

Towards the end of the interview, Berntsen offered a precise alternative to the growth paradigm, echoing the radical stance he took in 1994 at the first international conference on sustainable consumption:

> If the driving forces in the economy dictate that we must have growth, but that growth must be "green" to avoid environmental disaster, then we should take half of that growth to build our domestic society, and give the other half away to another part of the world.

Unfortunately, few of today's active national politicians from Norway's three largest parties—the Labour, Progress, or Conservative parties—share Berntsen's perspective. The "full breadbox" remains a taboo subject in the halls of power.

References

Aall, C. (2002). Framveksten av. det kommunale miljøvernet i Norge. Fra miljøvern til bærekraftig utvikling? [The emergence of municipal environmental protection in Norway]. In C. Aall, K. G. Høyer, & W. M. Lafferty (Eds.), Fra miljøvern til bærekraftig utvikling i kommunene. Erfaringer med Lokal Agenda 21 (pp. 29–70). Gyldendal akademisk.

Aall, C. (2011). The energy use of leisure consumption in Norway: Analysis and reduction strategy. *Journal of Sustainable Tourism,* 19(6), 729–745. https://doi.org/10.1080/09669582.2011.560440

[27] Lohne (2019, October 13).

Aall, C., Klepp, I. G., Støa, E., Engeset, A. B., & Skuland, S. (2011). Leisure and sustainable development in Norway: Part of the solution and the problem. *Leisure Studies, 30*(4), 453–476. https://doi.org/10.1080/02614367.2011.590521

Aall, C., Lafferty, W., & Bjørnæs, T. (1999). *Kartlegging av hindringer i prosjekt Bærekraftige lokalsamfunn* [Mapping barriers in the Sustainable Communities project] (Report 99:01 / TA 1607). Statens forurensingstilsyn (SFT).

Aall, C., & Norland, I. T. (2002). *Det økologiske fotavtrykk for Oslo kommune* [The ecological footprint for Oslo municipality] (VF-report 2/02). Western Norway Research Institute/ProSus.

Aall, C., & Norland, I. (2005). The use of the 'ecological footprint' in local politics and administration: Results and implications from Norway. *Local Environment, 10*(2), 159–172. https://doi.org/10.1080/1354983050004245З

Daxbeck, H., Kisliakova, A., & Obermosterer, R. (2001). *Der Ökologische Fussabdruck der Stadt Wien* [The ecological footprint of the city of Vienna]. Ressourcen Management Agentur.

De la Court, T., Schönbeck, H., & de Boer, M. (2000). *De Ecologische Voetafdruk van acht Nederlandse gemeenten* [The ecological footprint of eight Dutch municipalities]. De Kleine Aarde.

Hakanen, M. (1999). *Some Finnish ecological footprints at the local level.* Association of Finnish Local and Regional Authorities.

Hille, J. (1995). *Sustainable Norway: Probing the limits and choices from an environmental perspective.* Project for an Alternative Future.

Hille, J., Aall, C., & Klepp, I. G. (2007). *Miljøbelastninger fra norsk fritidsforbruk en kartlegging* [Environmental impacts of Norwegian leisure consumption—a survey] (VF-report 1/07). Western Norway Research Institute.

Hille, J., Storm, H., Sataøen, H., Aall, C., & Holden, E. (2008). *Miljøbelastningen fra norsk forbruk og norsk produksjon 1987–2007* [The environmental impact of Norwegian consumption and production 1987–2007] (VF-report 2/2008). Western Norway Research Institute.

Holden, M., & Norland, I. T. (2005). Three challenges for the compact city as a sustainable urban form: Household consumption of energy and transport in eight residential areas in the Greater Oslo Region. *Urban Studies, 42*(12), 2145–2166. https://doi.org/10.1080/00420980500332195

Høyer, K. G. (1997). Sustainable development. In D. Brune, D. Chapman, & M. Gwynne (Eds.), The global environment (pp. 1185–1208). VCH Publ.

Kommunenes Sentralforbund. (1993). Tenke globalt—handle lokalt. Lokalt prioriterte satsningsområder for miljøvernarbeidet [Think globally—act locally. Locally prioritized areas of focus for environmental protection work].

Lohne, J. (2019, October 13). *Thorbjørn Berntsen:—Jeg har for lengst mistet trua på det derre økonomisk-vekst-samfunnet. Vi har drekki og eti nok. Magasinet for fagorganiserte/ FriFagbevegelse.* https://frifagbevegelse.no/anbefaling/thorbjorn-berntsen%2D%2Djeg-har-for-lengst-mistet-trua-pa-det-derre-okonomiskvekstsamfunnet-vi-har-drekki-og-eti-nok-6.158.655440.603a3f29fd

Næss, A. (1976). Økologi, samfunn og livsstil [Ecology, society and lifestyle]. Universitetsforlaget.

Norwegian Government. (2008, March 28). *Nordmenns økologiske fotavtrykk er målt* [Press release]. Ministry of Children and Equality & Ministry of the Environment. https://www.regjeringen.no/no/dokumentarkiv/stoltenberg-ii/bld_2006-2009/Nyheter-og-pressemeldinger/pressemeldinger/2008/nordmenns-okologiske-fotavtrykk-er-malt-/id504420/

NOU 2006:18. (2006). *Et klimavennlig Norge* [A climate-friendly Norway] (NOU 2006:18). Miljøverndepartementet.

Rønnevik, J. S. (2019). *Approaching the limits: The environmental load of Norwegian leisure consumption* [Master's thesis, Western Norway University of Applied Sciences]. HVL Open.

Sabatier, P. A. (1986). Top-down and bottom-up approaches to implementation research: A critical analysis and suggested synthesis. *Journal of Public Policy, 6*(1), 21–48. https://doi.org/10.1017/S0143814X0000384X

SOU 2005:51. (2005). *Bilen, biffen, bostaden: Hållbara laster—smartare konsumtion* [The car, the steak, the housing: Sustainable burdens—smarter consumption]. Fritzes Offentliga Publikationer.

Stortinget. (1994, 3. mars). *Forhandlinger i Stortinget nr. 65 (1993–94). Debatt om Meld. St. nr. 4 (1992–93) Langtidsprogrammet 1994–97* [Proceedings of the Storting No. 65 (1993–94). Debate on White Paper No. 4 (1992–93) The Long-Term Programme 1994–97] (pp. 2921–3016).

Thompson, J. D. (1967). *Organizations in action: Social science bases of administrative theory.* McGraw-Hill.

Tiitu, M., Viinikka, A., Järvinen, L., & Viirret, E. (2019). Towards a comprehensive consumption-based carbon footprint of municipalities: A case study of Helsinki. *Sustainability, 11*(16), Article 4482. https://doi.org/10.3390/su11164482

Vedung, E. (1998). *Public policy and program evaluation.* Transaction Publishers.

Vindegg, M., Christensen, I., Aall, C., Arnslett, A., Tønnesen, A., Klemetsen, M., Temesgen, A. M., Hovelsrud, G. K., & Selseng, T. (2022). *Barrierer for klimatilpasning på lokalt og regionalt nivå* [Barriers to climate adaptation at local and regional levels] (M-2263). CICERO/Norwegian Environment Agency.

Wackernagel, M., & Rees, W. E. (1996). *Our ecological footprint: Reducing human impact on the earth.* New Society Publishers.

Chapter 4
Sustainable Tourism—From "Best Games Ever" to "Naive Technological Optimism"

Abstract This chapter analyzes the development of the concept of sustainability within tourism research and policy, using the 1994 Winter Olympics in Lillehammer as a starting point. The author describes how the Games marked the beginning of an ambitious environmental initiative that later became institutionalized through national strategies. Applying a framework that distinguishes between "hard" (technological) and "soft" (societal) strategies, the chapter argues that tourism policy has fallen into the trap of technological optimism. The chapter highlights the critical distinction between intensity problems (emissions per unit) and volume problems (total emissions), demonstrating that increased efficiency in the transport sector is often offset by increased travel activity. Using examples from Norwegian tourism strategy and global emissions data, the author concludes that a genuine sustainable transformation requires a shift from techno-centric solutions to political measures that dare to address and limit the total volume of long-distance travel.

Introduction

"You have presented to the entire world—the best Olympic Winter Games ever!" This was how the president of the International Olympic Committee (IOC), Juan Antonio Samaranch, summarized the conclusion of the Winter Olympics in Lillehammer in 1994. Beyond the sporting and audience success—which was likely what Samaranch had in mind—this was the first time environmental considerations were central to the planning and execution of the Olympic Games. The organizers placed great emphasis on providing efficient public transportation between event venues, implemented strict requirements for waste management and energy-efficient solutions at all facilities, and prioritized the use of local resources in venue construction and the Games' food profile. The commitment to an environmentally friendly Olympics in Lillehammer led the IOC to introduce a new environmental standard after the Games. In Norway, this initiative would prove important for the further development of environmental and, eventually, sustainability efforts in Norwegian tourism.[1]

---

[1] Teigland (1996).

C. Aall, *Sustainable Development*, https://doi.org/10.1007/978-3-032-35930-8_4

The World Commission's 1987 report does not mention tourism at all. The UN's action plan for sustainable development, Agenda 21, adopted at the Rio Conference in 1992, covered 40 topics and 115 programs. Despite its broad thematic scope and ambition to include all sectors of society, the tourism industry was only included through a few brief references to ecotourism as a tool to promote sustainable development.[2] Nevertheless, tourism was the first sector to follow the recommendations in Agenda 21 by developing its own Agenda 21, which was presented in 1995.

The UN Commission on Sustainable Development (CSD) addressed the limited attention given to tourism in Agenda 21. In 1999, the CSD selected tourism as the theme for its annual conference on the follow-up to Agenda 21. The conference resulted in a call for the further liberalization of international tourism regulations as a means to increase tourist flows to and between developing countries. According to the final document, this would strengthen the development aspect of sustainable development in the poorer parts of the world.[3] In other words, strengthening tourism was considered a tool to promote sustainable development in low-consumption countries, despite the obvious environmental consequences of increasing the scale of global tourism-related transport.[4]

This sustainability paradox—arising at the intersection of environmental and economic sustainability—would also come to characterize the Norwegian debate on sustainable tourism, a point I have raised repeatedly in various projects and public discussions, almost to the point of monotony. Another feature that became increasingly apparent as the Norwegian debate evolved was the belief in technological solutions to the greatest of all sustainability challenges facing the tourism industry: how to eliminate CO_2 emissions from travel itself.

While the official path to sustainability was being paved in the wake of Lillehammer, my own engagement with the theme began not in a boardroom, but at a dinner table in Sogn, where the inherent conflict between "travel" and "life" first became apparent.

Travel Less—Live More!

One of my first public appearances addressing sustainable tourism was at the National Rural Tourism Conference in Sogndal in 1996. My role was twofold: I was a speaker, but also part of the hosting team for the conference dinner. The dinner was held at Henjatunet—one of Norway's first full-scale organic, farm-to-table restaurants—where my wife was the chef. While she prepared the meal, I managed babysitting duties, though I fortunately managed to join the dinner.

---

[2] Johnson (1993).

[3] Commission on Sustainable Development (CSD) (1999).

[4] Aall and Teigland (2002).

Seeking to provoke a shift in perspective, I teamed up with Lone Lamark, a prominent advocate for sustainable tourism who was then working for the Norwegian Tourist Board (Norges Turistråd). I persuaded her to launch a provocative slogan: "Travel Less—Live More!"

In Norwegian, this was a deliberate, subversive play on the very word for tourism: *Reiseliv* (literally "travel-life"). By splitting the word, we challenged the industry to prioritize the *life* aspect—the good tourist experience—over the *travel* (which in modern tourism, in most cases, is reduced to a means to an end).

As I'd anticipated, the slogan met immediate friction. The next day, on a bus tour, I overheard a group of indignant participants. One passenger's voice stood out: "So we're supposed to stop traveling altogether? Is that what the Tourist Board thinks sustainable tourism means?"

This visceral reaction illustrates the early resistance to addressing the volume problem in tourism. Fast forward twenty years, and a new generation picked up the mantle. The Norwegian branch of Protect Our Winters (POW), in collaboration with the youth wing of the Norwegian Trekking Association (DNT ung), launched the "Environmental Code of Conduct" (*Miljøvettreglene*).

The initiative was a clever play on the iconic Norwegian "Mountain Code" (*Fjellvettreglene*)—the ten commandments of mountain safety that every Norwegian skier knows. My eldest son, who had recently founded the first international chapter of POW, authored this new code. The first rule reads: "Travel less, experience more!"

Emergence of a New Field of Research on Sustainable Tourism and the Launch of the Slogan "Travel Less—Live More!"

The provocative slogan "Travel Less—Live More!" was not a marketing stunt. It reflected a rigorous analytical shift being developed by a new research cluster. Six researchers helped establish the research area on tourism, environment, and sustainable development at Western Norway Research Institute (WNRI): Karl Georg Høyer, Erling Holden, Otto Andersen, Jon Teigland, Eivind Brendehaug, and me. Only one of the original group—Jon Teigland—had prior experience in tourism research. The rest of us brought expertise in sustainable development and transport into the field of tourism research. During the 2000s, WNRI secured several major research projects on sustainable tourism, which provided the foundation to double the number of staff in this area. As a result, in 2007, we hired Stefan Gössling, a highly cited researcher who published (and still publishes) internationally and prolifically in the field. Other researchers at WNRI who have since made significant contributions to the sustainable tourism literature include Hans Jakob Walnum, Agnes Brudvik Engeset, and Morten Simonsen.

Criticism of the Winter Olympics in Norway: Two Rounds of Resistance—One More to Come?

Before Jon Teigland joined WNRI, he had become a prominent—and perhaps "nationally disliked"—figure due to his rigorous critique of the 1994 Lillehammer Winter Olympics. Prior to the Games, he challenged the pervasive narrative from authorities and the business sector that the Olympics would trigger a golden age of growth for Norwegian tourism. After the event, Teigland demonstrated—first in a 1996 report and later in his doctoral dissertation (2000)—that the modest growth observed was largely coincidental, explained by Germany introducing an extra week of vacation in 1995.

The "Olympic legacy" was sobering: While 6,200 new beds were added in the region, the tourist influx failed to meet expectations. The state had invested NOK 200 million in the alpine resorts of Hafjell and Kvitfjell, yet both were eventually sold to new owners for a symbolic NOK 1 to avoid total bankruptcy. Within five years, 40% of the full-service hotels in Lillehammer had either gone bankrupt, been demolished, or been converted into student housing.

In 2013, the debate resurfaced when Oslo and Lillehammer applied for the 2022 Winter Olympics. Proponents recycled the same growth projections Teigland had debunked two decades earlier. Teigland and I issued a press release highlighting the Lillehammer failures. Predictably, it sparked significant controversy.

The following year, Cenk Demiroglu—a visiting researcher from Boğaziçi University—joined us; as part of his research stay, he calculated future snow reliability for the proposed 2022 venues. Our analysis, projected to the winter of 2022, indicated "very low reliability" for natural snow in Hafjell and Kvitfjell. The outlook for the Oslo venues was even bleaker. In an interview with *forskning.no*, I pointed out that over the previous nine years, February had averaged only four days cold enough to produce artificial snow.

Shortly thereafter, the Norwegian government withdrew the application. The 2022 Games were eventually awarded to Beijing, becoming the first Winter Olympics in history to rely almost 100% on artificial snow. This transition—from "The Best Games Ever" in Lillehammer to the "Dry Games" in Beijing—perfectly encapsulates the collision between tourism's growth imperative and the hard reality of the climate crisis.

Watching the Norwegian euphoria after the 2026 Winter Olympics in Italy—where Norway once again dominated the medal table—revealed a profound cultural paradox. While the 2022 Oslo bid was rejected partly due to environmental concerns and astronomical costs, the national mood seems to have shifted back toward a "Yes to the Olympics" sentiment. However, this raises a fundamental question of consistency.

If Norway continues to pour massive resources—both financial and human—into developing elite winter athletes to secure Olympic gold, it rings hollow to refuse to host the Games based on sustainability or cost. Opposing the Olympics while maintaining a state-sponsored "medal factory" ignores the systemic link between the two. You can't celebrate the output (the medals) while rejecting the infrastructure (the Games). If we truly believe that the environmental and economic costs of such mega-events are too high, we must also dare to question the very resources we spend on bringing forth the athletes who make these events possible. This points to a deeper barrier: the difficulty of scaling back a national passion even when it conflicts with our stated sustainability goals.

The failure of the "Olympic legacy" to produce sustainable growth serves as a macro-level warning: supply-side optimism rarely accounts for the complex reality of travel behavior. To understand why such grand visions often fail, we must look at the micro-level—specifically the entrenched holiday habits of the population.

The foundational tourism research expertise at WNRI was established through two related projects funded by the first tourism research program of the Research Council of Norway: One focused on factors influencing tourism demand,[5] and the other was an extensive panel study on Norwegians' holiday habits.[6] The latter highlighted the remarkable stability of holiday behavior and became widely cited in public reports and the media. At the same time, as indicated in the report, although Norwegians' holiday habits remained surprisingly stable over the period covered by the study (1986–1999), the data showed signs of impending changes in the form of increased international air travel for holidays—something that would indeed materialize fully.

The first major tourism project at WNRI to address sustainability and the environment explicitly began in 1995. I played a central role in the project. The title of the final report—"Sustainable Tourism: Much 'Tourism' and Little 'Travel'"—eventually became a catchphrase for myself and, by extension, the Institute's position in the debate on sustainable tourism. The catchphrase originated from a framework developed by my two colleagues, Karl Georg Høyer and Morten Simonsen, in a commission for the Ministry of Industry and Energy.[7]

The framework distinguishes between two main types of tourism activities—*stationary* (such as accommodation and restaurants) and *mobile* (primarily travel to and from the destination)—as well as two main types of environmental problems.

---

[5] Teigland (1997).

[6] Teigland (2000).

[7] Høyer and Simonsen (1996).

Intensity problems are associated with local challenges resulting from a concentration of tourism activities. These can lead to exceeding local carrying capacities, for example, high concentrations of air pollution in cities from tourism-related transport. *Volume problems* arise from many small impacts that together have significant global effects. Examples include greenhouse gas emissions from tourism-related transport. The "Sustainability Paradox" lies in this: An industry can be locally "green" (intensity) while being globally "black" (volume) due to emissions from travel itself.

In our subsequent analyses, we pointed out that until the mid-1990s, attention had focused primarily on intensity-related issues stemming from stationary tourism activities. I often illustrate this by referring to the small signs we have all seen in hotel bathrooms, urging us to "save the world by hanging up your towel to avoid unnecessary laundering". The challenge for the tourism industry, if it truly wishes to contribute to "saving the world", is to shift attention to the volume-related issues associated with the mobile aspects of tourism.

In the article "Sustainable Tourism or Sustainable Mobility?" Karl Georg Høyer (see Footnote 25) noted that tourism activities marketed as environmentally friendly—under headings such as "ecotourism," "green tourism," "sustainable tourism," and similar—often involve longer distances to more remote locations and more frequent use of air travel and private transport than conventional forms of tourism. He, therefore, emphasized that the priority for aligning tourism with the sustainability goal is to reduce mobility, especially air travel. Furthermore, Høyer pointed out that measures aimed at increasing profitability in tourism—an obvious priority for any business—do not in themselves improve the sustainability of tourism, even when presented under headings such as "economic sustainability".

About Sawing Off the Branch You're Sitting On—With a Chainsaw!

Norway's appeal as a holiday destination is deeply rooted in the image of vast, pristine nature combined with vibrant local communities and small-scale agriculture that preserves the historical cultural landscape. In 2004, the Norwegian fjords—led by the Sognefjord—were ranked by *National Geographic Traveler* magazine as the world's best-preserved and most beautiful travel destination. Yet, from time to time, aggressive tourism development proposals emerge in my beautiful fjord county that run directly counter to these very qualities.

In 1999, I unexpectedly found myself at the center of one such proposal. At the time, our family was renting a romantic little cabin in Veitastrond, a secluded farming valley in Sogn that perfectly embodies the Norwegian rural ideal. The cabin was originally a *sel*—a traditional stone or timber hut located on a mountain farm (*støl*). Historically, these buildings served as seasonal dwellings for dairymaids and as workshops for making cheese over an open fire, but this particular one had been carefully converted into a holiday cottage.

One day, our landlord, Asbjørn Nes, stopped by after milking his goats in the summer barn next to our cabin. He presented his vision: a restaurant on the mountain peak Kaldakari, inside Jostedalsbreen National Park, offering spectacular glacier views.

The business concept involved transporting guests by helicopter from a departure point right outside the cabin up to the peak. The restaurant would specialize in local and organic food. Then, he asked if my wife would be interested in the position of head chef. By then, she had become known as one of the county's finest chefs, with experience from farm-to-table restaurants and as an author of cookbooks on organic and local cuisine. She politely declined, and I—perhaps somewhat recklessly—bet a case of beer that he would never receive government approval for such a venture.

However, it would take six years and involvement from three different Ministers of the Environment before the state finally mustered the courage to reject the application. The process involved heated media debates, including a national TV appearance on NRK, where Asbjørn Nes and I were interviewed as the opposing voices. I'm still friends with Asbjørn today, but I have yet to receive my case of beer.

Sustainable Tourism Becomes Part of Norwegian Tourism Policy

Today, the authorities promote "sustainable tourism" as the overarching goal for Norwegian tourism development. The journey was long and uneven, and I am uncertain whether the development may have taken a wrong turn at one of the crossroads along the way. I will make three stops along this road:

- The first national tourism strategy to place sustainability high on the agenda.
- A national set of indicators for sustainable tourism.
- The world's first national system for sustainability certification of destinations.

I will briefly reflect on the use of these systems and how, despite initial promises, they have fallen short of creating a truly sustainable tourism sector. Finally, I will briefly looking back at one initiative from the Lillehammer Olympics—focused on environmentally friendly and locally sourced food—as a rare, genuinely successful outcome from this era.

National Tourism Strategy from 2006

The development of a Norwegian tourism policy aligned with the goal of sustainable development initially took a conceptual detour via the term "quality tourism," focusing on tourism activities related to mountain areas, particularly national parks. This was examined through a report for the Ministry of the Environment by WNRI titled "Commercial Potential for Quality Tourism,"[8] alongside an international literature review that also introduced the idea of establishing a national institute for quality tourism at WNRI.[9] Eventually, however, the concept of sustainability prevailed, and in 2006, sustainable development was adopted as one of three overarching goals for Norwegian tourism development through the government's tourism strategy "Valuable Experiences" presented in 2006.[10]

The government provided two reasons for emphasizing the goal of sustainable tourism. The *first* was to fulfill international political commitments regarding the environment and climate. The strategy even went so far as to introduce the goal of a CO_2-neutral tourism sector, but without identifying specific measures or clearly defining what was meant by "sustainable tourism." The *second* was to preserve natural and cultural landscapes as Norway's most important sources of experiences. My colleague Stefan Gössling, drawing on his broad international perspective, has pointed out that the scope of Norway's commitment to sustainable tourism at this time was unique.[11] Nevertheless, in the public debate on sustainable tourism, the research group from WNRI—and, if I may say so, myself in particular—adopted a critical stance toward the follow-up of the tourism white paper. This national ambition of CO_2-neutrality frequently clashed with the pragmatic—and often cynical—realities of local commercial interests.

"Local Sustainability is Good—Global Sustainability is Someone Else's Problem."

In January 2001, I was invited by the municipality of Tinn to address a public meeting regarding large-scale tourism development around Gaustatoppen. This is no ordinary mountain. Standing at 1883 m, its distinctive pyramidal shape has made it a national icon. On a clear day, the summit offers a view of one-sixth of mainland Norway. Historically, it was a central motif for the nineteenth-century National Romantic painters, and it holds a sacred place in Norwegian *friluftsliv* (outdoor life). When the Norwegian Trekking Association (DNT) was founded in 1868, Gaustatoppen was one of its first priorities, leading to the construction of a mountain cabin in 1893—a move intended to "democratize" the wilderness.

---

[8] Aall et al. (2003).

[9] Aall and Groven (2004).

[10] Nærings-og handelsdepartementet (2006).

[11] Gössling et al. (2012).

True to form, I concluded my presentation by advocating for the maxim "Travel Less—Live More!" I recommended that the municipality not only look inward but also adopt ambitions to reduce the global environmental impact caused by greenhouse gas emissions from travel to and from the region.

The prominent real estate developer and "hotel king" Arthur Buchardt responded with striking bluntness. He suggested that while local sustainability was a municipal responsibility, global issues were not. In his view, the municipality should focus exclusively on local jobs and nature conservation, and—in his own unvarnished words—"not give a damn" about global sustainability. That, he said, was a burden for national and international politicians to carry.

At the time, Tinn municipality had officially committed to sustainable tourism through the Local Agenda 21 framework. Nevertheless, subsequent development proceeded largely in line with the "hotel king's" vision. As of this writing, the neighboring municipality of Hjartdal is debating a plan for 1900 new holiday homes in the same area. In the annual "Nature Battle" (*Naturkampen*)[12] ranking, which assesses how Norway's 357 municipalities protect their natural assets, Tinn and Hjartdal rank poorly at 218th and 315th, respectively.

Ten Principles for Sustainable Tourism—Without Reference to "Climate"

As mandated by the national tourism strategy, Innovation Norway conducted an extensive process involving participants from industry, public administration, and research—including us from WNRI—to develop what was initially called a national set of indicators. The process eventually resulted in a set of principles describing the content of efforts to promote sustainable tourism. The work culminated in the formulation of ten principles, structured in a manner similar to the UN World Tourism Organization's distinction between economic, social, and environmental sustainability (see Table 4.1). I have on several occasions criticized the approach chosen by Innovation Norway and parts of the tourism sector, both in connection with the process of developing the indicators and later in the public debate on sustainable tourism—evidently to the considerable irritation of both Innovation Norway and others in the tourism industry. My criticism concerned two main issues: the understanding and strong emphasis on *economic* sustainability, and the understanding and insufficient emphasis on *ecological* sustainability.

[12] Sabima (n.d.).

Table 4.1 Innovation Norway's ten principles for sustainable tourism[13]

Conservation of nature, culture, and environment	Strengthening of social values	Economic viability
(1) *Cultural richness*—To respect, further develop, and highlight the local community's historical cultural heritage, authentic culture, traditions, and distinctive character (2) *Physical and visual integrity of the landscape*—To preserve and enhance landscape qualities, both in urban and rural areas, so that the physical and visual integrity of the landscape is not degraded (3) *Biological diversity*—To support the conservation of natural areas, wildlife, and habitats, and to minimize their destruction (4) *Clean environment and resource efficiency*—To minimize pollution of air, water, and land (including noise) by tourism enterprises and tourists, as well as to minimize their generation of waste and consumption of scarce and non-renewable resources	(5) *Local quality of life and social values*—To preserve and strengthen the quality of life in the local community, including social structures, access to resources, facilities, and public goods for all, and to avoid any form of social degradation and exploitation (6) *Local control and engagement*—To involve and empower the local community and local stakeholders in planning, decision-making, and the development of local tourism (7) *Job quality for tourism employees*—To improve the quality of tourism jobs (both direct and indirect), including wage levels and working conditions without discrimination based on gender, race, disabilities, or other factors (8) *Guest satisfaction and safety; quality of experience*—To ensure safe, satisfying, and enriching tourism destinations through local value creation	(9) *Economically viable and competitive*—To ensure the viability and competitiveness of tourism destinations in a long-term perspective by maximizing tourism's value creation in the local community, including the value tourists leave locally (10) *Economically viable and competitive tourism enterprises*—To ensure the viability and competitiveness of tourism enterprises from a long-term perspective

My first criticism concerned the relevance—and subsequently the emphasis—placed on including principles of economic profitability for the tourism industry. Ensuring the financial bottom line for a business and maintaining employment levels in a local community are, of course, legitimate goals and considerations. However, how relevant is it to include such objectives as part of sustainability efforts in one of the world's wealthiest countries with the highest standard of living, lowest unemployment, and least social inequality? My point was that economic sustainability is primarily about meeting basic needs and reducing economic disparities between

[13] Innovasjon Norge (2011).

and within countries. To the extent that considerations of economic sustainability are relevant in affluent countries like Norway, such considerations must be viewed as *means* to achieve ecological sustainability, not as ends in themselves. In this context, I warned that including principles of economic sustainability—without simultaneously emphasizing that such goals in a Norwegian context must be subordinated to ecological sustainability—would lead tourism development into a *sustainability paradox*, where weakening ecological sustainability is accepted if economic sustainability is strengthened. This warning would eventually prove justified.

My *second* criticism concerned the absence of a principle addressing *climate*. Innovation Norway responded to the criticism by pointing out that the reduction of greenhouse gas emissions was included in the fourth principle: "Clean environment and resource efficiency." However, given that greenhouse gas emissions represent by far the greatest sustainability challenge facing the tourism industry, it seems at best peculiar that the word "climate" does not even appear in the explanation of the principle supposedly meant to address the tourism sector's sustainability challenge. I also criticized Innovation Norway for omitting the issue of climate change adaptation, noting that the government, one year earlier, had presented a government green paper from 2010 on climate change adaptation,[14] which included a dedicated chapter highlighting the need for the Norwegian tourism industry to prepare for a changing climate.

Naturally, I was by no means alone in voicing these concerns. Many dedicated individuals within the tourism industry, particularly from the small-scale sector focused on nature-based tourism and farm tourism, have raised similar concerns. The influential Norwegian travel journalist Odd Roar Lange, who runs the online newspaper *The Travel Inspector*, has recently joined the list of critics. In an article published on his website dated 18 February 2022, he refers to the ten sustainability principles as "The Great Norwegian Sustainability Hoax".[15] None of these criticisms has to date prompted Innovation Norway to make any changes to its original framework since 2011.

The World's Most Advanced System for Sustainable Tourism Certification

The most clearly innovative follow-up to the government's tourism strategy from 2007 was the development of a certification system for sustainable destinations. In an international context, this is a groundbreaking instrument whereby national authorities establish a labeling scheme for sustainable tourism that goes beyond individual products (such as eco-labeled food items) and individual producers (such as eco-labeled hotels) to encompass an entire destination. The scheme also requires

[14] NOU 2010:10.

[15] The Travel Inspector (2022, February 18).

Table 4.2 Criteria and indicators related to "climate" in the Norwegian system for sustainable destination certification

Criteria	Indicators
The destination management has set goals to reduce both local greenhouse gas emissions and emissions related to guests' transportation to, from, and within the destination	• Does the municipality have a climate plan with goals, measures, and instruments for reducing greenhouse gases within the municipality's boundaries? Has the destination management set concrete, long-term goals for reducing emissions related to guests' transportation to, from, and within the destination, and for increasing the use of non-motorized travel? • CO_2 emissions from guests' transportation to and from the destination at the most recent available measurement. • Number of measures implemented by the destination management that reduce greenhouse gas emissions related to guests' transportation to, from, and within the destination in the past three years. • Number of measures implemented in the past three years that contribute to reducing greenhouse gas emissions from member businesses.
The destination management has identified risks and measures related to the tourism sector's adaptation to a changing climate	• Are local challenges and opportunities related to climate change identified? • Does the municipality (or municipalities) have a climate change adaptation strategy that includes outdoor recreation areas and/or tourism facilities? • Are climate change issues, including associated risks and considerations related to tourism, communicated to tourism businesses and guests?

that the tourism industry enter a binding partnership with local authorities. The system includes several supplementary tools, including an calculator that estimates greenhouse gas emissions from tourists' transportation to and from Norway, from the point of arrival in Norway to and from the destination, and within the destination.

To become certified, the destination must meet comprehensive requirements. The current version of the standard contains 31 criteria and 89 indicators distributed across five areas: destination management, strategic foundation, conservation of nature, culture, and environment, strengthening of social values, and economic viability. In contrast to Innovation Norway's ten principles for sustainable tourism (see Table 4.1), the system for the sustainability certification of destinations includes a range of requirements and indicators related to climate, including requirements for reducing greenhouse gas emissions to and from the destination and requirements addressing the need for climate change adaptation. In 2024, an addendum was introduced to the standard for destinations with cruise ship calls, which included the following four additional climate indicators: (1) Have the local port(s) implemented the Environmental Port Index (EPI); (2) Average EPI score for cruise ship calls in the previous calendar year; (3) CO_2 emissions from cruise ship calls to/from and at the port(s) at the destination in the previous calendar year; and (4) Proportion of cruise ships that used shore power at the port(s) at the destination in the previous calendar year. Indicator (3) is particularly relevant in light of the public debate on environmental requirements for cruise tourism, which is most often reduced to a demand for "shore power"—that is, that cruise ships must disconnect their fossil fuel-based generators and connect to (and pay a premium for) shore-based power facilities (presumably supplied by renewable energy sources). My colleagues have calculated that emissions from all port stays on average account for 10–15% of the total emissions from the entire voyage.[16] Thus, indicator (3) also captures the remaining majority of emissions (Table 4.2).

Between 2009 and 2013, the system was developed and tested in dialogue with four pilot destinations—Trysil, Røros, Vega, and Lærdal—and has since 2014 been a public system that currently includes approximately 50 destinations.[17] Our analysis of the four pilot destinations showed that they prioritized local sustainability over global sustainability. The focus was placed on measures such as improved waste management, better public transportation, and strengthening cooperation between agriculture and tourism.[18] A master's thesis at the Norwegian University of Life Sciences reported similar findings, highlighting that Trysil met only half of the principles, despite being approved under the certification scheme.[19] At the time of writing, a one-year project has been initiated to evaluate the scheme.

Destination Trysil—The Quintessential Example of Innovation Norway's Sustainability Paradox.

Destination Trysil markets itself as Norway's premier ski and mountain biking destination. It boasts over 1,000,000 guest nights, 30,000 beds, and an extensive network of 69 ski slopes and 65 km of purpose-built cycling trails. In 2024, international visitors generated 85% of all commercial guest nights.

Trysil was one of four pilot projects chosen by Innovation Norway to develop and test the national certification scheme for "Sustainable Destinations." During the pilot, the destination formulated an ambitious vision: to become the Nordic region's leader in sustainability. Their strategy focused heavily on enhancing economic viability while addressing environmental concerns

[16] Simonsen et al. (2018).

[17] Innovasjon Norge (2024).

[18] Brendehaug (2016).

[19] Sogn (2014).

through local measures, such as improving public transport and implementing energy efficiency initiatives in hotels and ski resorts.

But here's the catch. While this sustainability work was underway, Destination Trysil was simultaneously collaborating with the Swedish resort of Sälen to build a new international airport. Branded as "The Scandinavian Mountains Airport," it was designed specifically to serve the ski resorts of Trysil and Sälen, offering direct flight routes from markets such as England, Belgium, the Netherlands, Germany, and Denmark. The projections—or rather, the optimistic expectations—were 60,000 passengers in the first year of operation (2020), with anticipated annual growth of up to 40,000 passengers. These numbers have yet to materialize; in 2024, the passenger count reached just 23,000.

On several occasions, I have pointed out the glaring sustainability paradox of pursuing a prestigious "sustainable" certification while actively working to increase the volume of high-emission air travel. I have addressed this criticism directly to representatives from both Destination Trysil and Innovation Norway. Their responses are telling: They either claim to see no paradox at all, or they simply shrug, suggesting that this inherent contradiction is just an unavoidable reality of modern tourism.

Emissions from Travel to and from Norway Are Finally Accounted for by the Tourism Industry

Throughout the 2000s, the debate on sustainable tourism developed much the same way as the broader sustainability debate in society, focusing increasingly on "climate." In 2017, Norway received its second white paper on tourism. Even eight years after Innovation Norway established a certification system for sustainable destinations—which included several requirements to reduce greenhouse gas emissions associated with travel to and from Norway—the government still refrains from prioritizing this work, as indicated by the following reference (p. 72)[20]:

> Globally, there is ongoing debate about whether and how transportation should be included in an assessment of the sustainability of tourism.

At the same time, the report highlights the focus on electric cars, electric buses, and hydrogen buses as an important part of the tourism sector's sustainability efforts. The report also notes that energy efficiency improvements in the aviation industry have reduced greenhouse gas emissions per passenger-kilometer, but at the same time acknowledges (p. 73):[21]

[20] Næringsdepartementet (2017).

> Increased traffic has nevertheless led to a net rise in greenhouse gas emissions from aviation.

Two years later—in 2019—NHO Reiseliv (the Confederation of Norwegian Enterprise's tourism division), presented a greenhouse gas inventory for Norwegian tourism prepared by the consultancy firm Stakeholder.[21] In contrast to the government's reluctance to include emissions from travel in the assessment of tourism's sustainability, NHO specifically sought to highlight this aspect. Unsurprisingly, the report demonstrates that greenhouse gas emissions from travel overwhelmingly dominate the tourism sector's climate footprint, with emissions from air travel constituting the largest single source, accounting for just over half of total emissions. According to the report, cruise emissions rank second, representing approximately 16 percent of the tourism sector's total greenhouse gas emissions. Government authorities have not subsequently produced an independent "official" greenhouse gas inventory for Norwegian tourism. Instead, they refer to this report.

NHO Tourism Issues a Counter-Statement: The Battle Over Emission Boundaries

In 2019, I was invited by the Norwegian Mountain Museum in Lom to help plan a national conference on sustainable tourism—the "Lom Conference." I had the honor of giving the opening keynote, where I presented new calculations of greenhouse gas emissions associated with Norwegian tourism. My analysis took a lifecycle approach, including emissions from international tourists traveling to and from Norway, as well as Norwegians' holiday trips abroad. This approach quite literally exceeded the boundaries of the official Norwegian greenhouse gas inventory, which only accounts for emissions that physically occur within the country's borders.

In my presentation, I demonstrated that total emissions from Norwegian-related tourism were roughly equivalent to the largest single source in the official national inventory: emissions originating from the production of Norwegian oil and gas. The media, of course, found this comparison irresistible. Predictably, my figures sparked strong protests and near-ridicule from parts of the tourism industry, particularly from NHO Reiseliv.

Three months later—in August 2019—NHO Reiseliv presented its own set of figures, produced by the consultancy firm Stakeholder. The motivation behind the report was unusually explicit. The introduction stated:

> This assignment was commissioned as a reaction to a presentation given by Aall at a tourism conference in Lom.

The figures produced by Stakeholder were remarkably similar to mine, with one crucial exception: their accounting included only Norwegians' flights *out of* Norway and foreign tourists' flights *into* Norway. By counting only one-way trips, they effectively halved the total emissions. Stefan Gössling and I pointed out this peculiarity in an op-ed published by the Research Council of Norway's

[21] Bjørndal and Thompson (2019, November 22).

news platform.[22] The consultants' response offered little clarity. They argued that the omitted emissions "belonged" to other countries. Nevertheless, a significant milestone had been reached: The major players in Norwegian tourism had finally accepted that emissions from travel to and from Norway must be part of the conversation.

The report from the consultancy firm Stakeholder also includes data on value creation, which the authors used to calculate emissions per krone spent in Norway from various market segments. Unsurprisingly, cruise tourists performed very poorly due to the combination of low spending per guest in Norway and very high emissions from travel to and from Norway, even though the distances were relatively short. The most common departure ports for cruises to Norway are Southampton in the United Kingdom and the German ports of Hamburg, Kiel, and Bremerhaven. The calculations showed that even long-haul (intercontinental) air tourists can have a better outcome than cruise tourists and other short-haul forms of tourism, as long as their economic spending in Norway is sufficiently high. For example, tourists from the USA, Canada, and China (with 16–17 kg CO_2 per overnight stay) performed better than short-haul tourists from Spain and Poland (with 19 and 24 kg CO_2 per overnight stay, respectively). In other words, this is another example of what I have previously called Innovation Norway's sustainability paradox: improved social sustainability measured in kroner per unit of greenhouse gas emissions comes at the expense of poorer environmental sustainability measured as the total amount of greenhouse gas emissions.

Based on these calculations, NHO Tourism concluded that "we must limit cruise traffic because emissions from this form of tourism are too high relative to the value creation in Norway", while also stating that "the Asian [air] market is the most important growth market," since according to the organization's metrics, higher value creation in Norway per tourist compensates for the high emissions from air travel.[23]

Cruise Becomes the Climate Scapegoat

The only truly new political governance directive concerning sustainability in the government's 2017 white paper on tourism pertained to the cruise industry. Although the authorities now officially acknowledged that greenhouse gas emissions from air travel were by far the largest source within Norwegian tourism, the climate focus was directed toward the cruise industry (p. 74) (see Footnote 20):

> [The government will] facilitate that local authorities, in collaboration with relevant stakeholders, work to ensure that the cruise industry reduces greenhouse gas emissions.

---

[22] Aall and Gössling (2019, November 21).

[23] NHO Reiseliv (2019, August 28).

An important aspect of this work has been extensive government financial support for the establishment of shore-power facilities, which allow cruise ships to shut down their fossil fuel engines and use electricity from the shore to operate all onboard systems while docked. In 2018, thirteen cruise ports in Norway joined together in a joint declaration with the goal of establishing shore-power at all Norwegian cruise destinations by 2025. As of 2020, there are plans to develop shore-power facilities in 20 of Norway's 32 main network ports, but only a few of these (Oslo, Kristiansand, Bergen, Ålesund) possess or are in the process of acquiring shore-power facilities for cruise ships.[24]

WNRI was among the first to calculate emissions from cruise tourism, and we quickly became known for characterizing cruises as a "climate offender." This is accurate, but only from an "intensity perspective" (see the typology in Fig. 4.1). When understood as a volume issue, and if we limit the sustainability perspective to greenhouse gas emissions, cruises account for a relatively small share compared to the far larger source: air travel. From an intensity perspective, however, cruise tourism is clearly at the top (see Fig. 4.2).

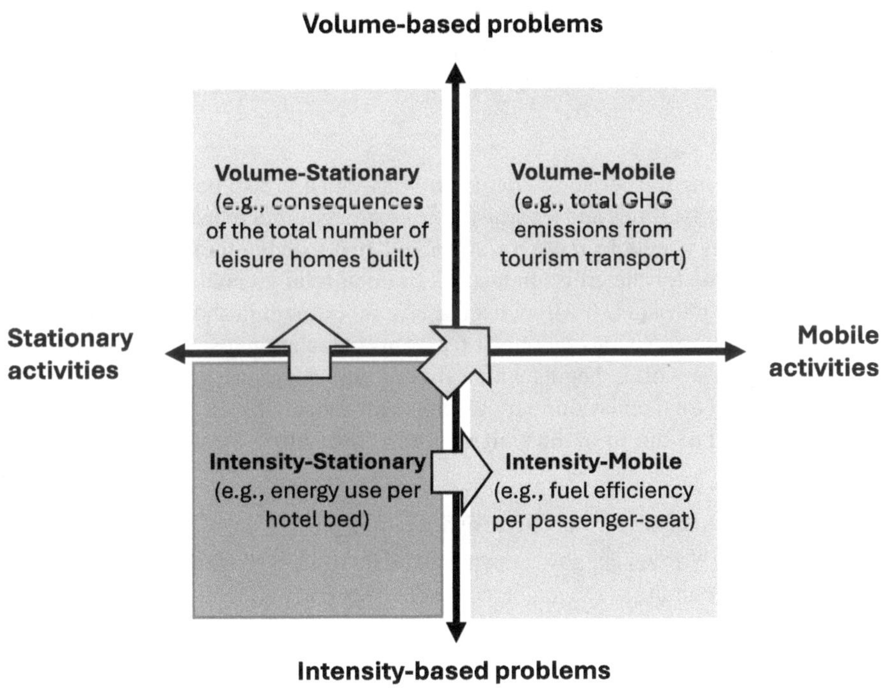

Fig. 4.1 Framework for describing the sustainability challenges of tourism, indicating the dominant understanding of sustainable tourism in the 1990s (intensity × stationary) and highlighting the need for an expanded understanding[25]

[24] Grønt Skipsfartsprogram (2020).

[25] Adapted from: Høyer (2000).

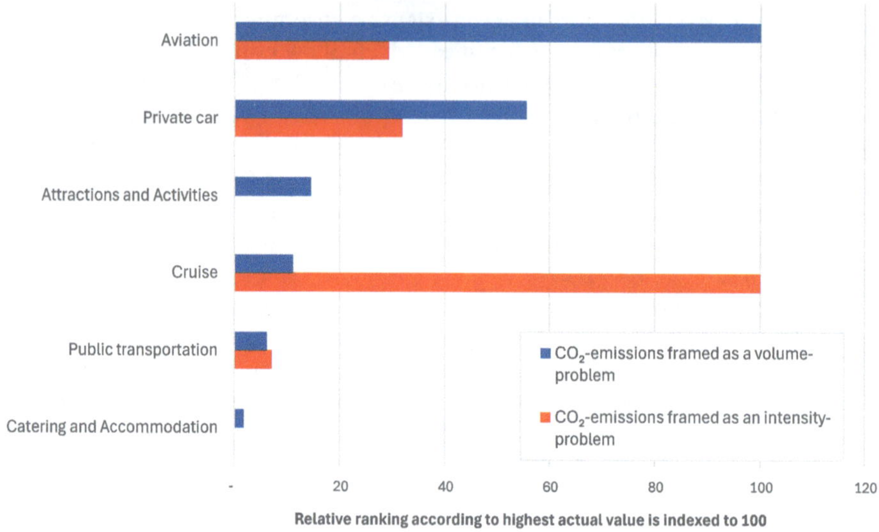

Fig. 4.2 Dual nature of tourism's climate impact: an intensity-based perspective (emissions per activity) versus a volume-based perspective (total global emissions per activity). Data derived from the following sources: United Nations World Tourism Organization (UNWTO) (2016/2019) and Sun et al. (2022)[26]

Our calculations further show that if all Norwegian cruise ports were equipped with shore-power facilities and all cruise ships were required to use them, total emissions from cruises would decrease by only about 10% (see Footnote 16). Using NHO Reiseliv's estimate for the cruise industry's share of total greenhouse gas emissions from Norwegian tourism (16%), such a massive investment in shore power facilities would reduce total emissions by only 1.6%. This calculation also does not consider that a cruise journey often begins with a flight to the departure port. Some calculations suggest that total emissions per passenger-km can double in certain cases when we include travel to and from the embarkation port by air.[27]

From "Cruise Destroyer" to "Cruise Saviour"

After both the Norwegian government and NHO Reiseliv began framing cruise tourism as a "climate scapegoat", the media frequently sought me out as a key

[26] Data derived from the following sources: United Nations World Tourism Organization (UNWTO) (2019) and Sun et al. (2022).

[27] Walnum (2011).

witness to the industry's environmental toll. My role included co-planning and appearing in the 2017 NRK documentary *Shit O'hoi*,[28] which critically examined the cruise sector's impact, as well as participating in an 2019 national election broadcast from the cruise port of Flåm,[29] where the debate centered on the conflict between cruise growth and environmental protection.

Then came a fascinating shift. The cruise industry gradually realized that my research also identified long-haul air travel as the true climate culprit in terms of total emissions volume. My message—once seen as a threat—was suddenly reframed as a defensive shield. So, they welcomed me into the fold and invited me to address the sector directly.

This included a keynote at the 2022 National Cruise Tourism Conference and a prominent event during Arendalsuka 2023. The latter, organized by the Bergen Chamber of Commerce and other regional players, had an evocative title: "Can Cruise Tourism Become the New Stockfish?"—a bold attempt to link a controversial modern industry to the historical backbone of the Norwegian economy.

The Struggle Over the Tourism Industry's Dependence on Air Travel

The remarkable global growth in tourism since the 1950s has been made possible by a dramatic increase in the availability of air travel at increasingly affordable prices. This growth began with the establishment of national flights, followed by routes within continents—primarily in Europe and the United States—and ultimately the development of intercontinental routes. The emergence of low-cost carriers in the early 2000s provided an additional boost to this trend. In Norway, this development is reflected in a quantum leap in Norwegians' international air travel outside traditional charter trips, as documented in a comprehensive survey of Norwegian consumption conducted by WNRI on behalf of the Ministry of Children and Equality and the Ministry of the Environment, in connection with the 20th anniversary of the publication of the World Commission's report (see Table 4.3). Subsequent studies have confirmed this pattern. In 2019, only five countries—the island states of Iceland, Malta, Seychelles, Cyprus, and the Maldives—had more air travel per capita than Norway.[30]

---

[28] NRK (2017, October 23).

[29] NRK (2019, August 21).

[30] Our World in Data. (n.d.).

Table 4.3 Development of Norwegians' passenger transport activity 1985–2005, passenger-kilometers and percentages[31]

Year	Railway and bus	Passenger car	Domestic flights	Charter flights abroad	Scheduled flights abroad
1985	7021	35,169	685	5508	1308
1997	7015	41,196	1488	5600	3555
2005	7449	49,488	1981	7524	12,602
Development 1985–2005	+6%	+41%	+189%	+37%	+863%

In an op-ed published in the journal *Samferdsel*—titled "Technological Measures: Not Enough to Reduce Greenhouse Gas Emissions from Aviation"—we took as our starting point the vision of the International Air Transport Association (IATA) for "zero-emission flights" by 2050.[32] We pointed out three technological measures to reduce greenhouse gas emissions from air transport: improved air traffic management, transition to new engines and aircraft types, and replacing fossil fuels with alternative energy sources (while largely retaining current aircraft designs). According to the IATA, improved air traffic management can reduce emissions by a maximum of 12%, while a realistic estimate of annual energy efficiency improvements of 1% per year—resulting from new engines and aircraft types—could theoretically reduce average fuel consumption per seat-kilometer by 33% from 2000 to 2040.[33] In the article, we noted that the aviation and tourism industries expect global air passenger numbers to grow by around 5% annually in the coming decades, with travel distances increasing even more. Historically, the combined emissions reductions from technological measures have been 1–1.5% per year, and we cited research suggesting this rate will continue, possibly with a declining trend. Subsequent research has confirmed this perspective.[34] We concluded our 2008 article by predicting that greenhouse gas emissions would continue to rise—by an estimated 3% annually (a long-term average that holds true even across the pandemic years)—and pointed out that it is impossible to envision how the IATA's vision of zero emissions by 2050 could be achieved through technological measures alone. For the aviation industry to meet this target, the volume of air travel must first be drastically reduced. Second, the nature of air travel must change to occur more slowly (not faster, as has been the trend throughout aviation history) and with smaller aircraft (not larger, which has also been the main trend). Overall, these changes will increase the cost of air travel, which in itself will reduce volume.

[31] Hille et al. (2008).

[32] Gössling et al. (2008).

[33] Peeters et al. (2005).

[34] Peeters et al. (2016), Gössling and Lyle (2021) and Gössling et al. (2024).

Avinor's Two Voices in the Debate on Aviation and Climate

Avinor, the state-owned company that operates most Norwegian airports, sits at the heart of the aviation–climate paradox. Their long-term forecasts predict a near doubling of departures from Norwegian airports between 2016 and 2040—growth that would necessitate a third runway at Oslo Airport, Gardermoen, by 2030.

In 2018, the Avinor Board sent an unusually candid letter to the Ministry of Transport in connection with the government's White Paper on State Ownership to the Storting. The board pointed out a fundamental contradiction: the societal mandate they had received—to facilitate growth in air traffic—was in direct conflict with Norway's national climate goals and international commitments.

Yet, while expressing these internal concerns, Avinor has been a vocal proponent of electric aircraft as a primary climate solution. A year after the board's letter, this strategy quite literally crash-landed. On 14 August 2019, an experimental electric Pipistrel Alpha Electro, purchased by Avinor, lost engine power on its way to Arendal. With Avinor's CEO, Dag Falk-Petersen, at the controls, the aircraft was forced into an emergency landing. It ended up upside down in Lake Nornestjønn. Fortunately, the pilot emerged uninjured.

Despite this setback, Avinor and the government continued to champion electrification as the future of the short-haul network. The stark contrast between rhetoric and reality was perfectly captured in 2025. In March, the airline Norwegian announced the purchase of ten conventional Boeing 737–800 aircraft, stating that the deal "fits well with our long-term fleet strategy".[35]

Only five months later, in August 2025, Avinor celebrated the first test flight of an electric scheduled aircraft between Stavanger and Bergen—a BETA Technologies ALIA model with room for just five passengers. The Minister of Transport, Jon-Ivar Nygård, declared in an interview: "This is the future".[36] I suspect he is absolutely right—though perhaps in a way he did not intend: a future where the "green solutions" are marginal while the core business remains firmly committed to fossil-fuelled expansion.

[35] Norwegian (2025, March 3).

[36] Avinor (2025, August 8).

Research in Support of Sustainable and Climate-Friendly Tourism

The Sustainable Destination Norway 2025 (SDN) project, which spanned the period 2007–2011, was one of three research projects funded as part of the government's follow-up to the 2007 tourism white paper. The project gave WNRI a public role in conducting research "for" sustainable tourism. SDN comprised three main activities:

- Organization of three consecutive national dialogue conferences on sustainable tourism ("Balestrand Summit" in 2008, 2009, and 2010).
- Development of a scenario model with an associated PC-based game for sustainable tourism development in Norway.
- Analyses of the sustainability profile of food services in Norwegian tourism.

The three dialogue conferences—the Balestrand Summit—became something of a national event, where we managed to gather around 100 participants each year, including representatives from both large and small tourism enterprises in Norway, national and regional tourism authorities, and, alongside us at WNRI, some of the leading and most-published international researchers in sustainable tourism. These included Daniel Scott, David Weaver, Bernhard Lande, Jean-Paul Ceron, Michael Hall, Paul Peeters, and Xavier Font. These individuals served on an international scientific advisory board for what we would eventually launch as a National Knowledge Centre for Sustainable Tourism.

Using Market Power to Drive Change: The Case of the Kviknes Hotel

The Balestrand Summit was held at the historic Kviknes Hotel, a magnificent example of the "Swiss-style" wooden architecture that graced Norway's fjords at the turn of the twentieth century. When we chose Kviknes as the venue for our three annual conferences, we made one non-negotiable demand: the catering had to reflect a genuine sustainability profile. This meant a commitment to organic, locally sourced ingredients and high-quality vegetarian options—a concept still relatively novel in the Norwegian hotel industry at the time. After some initial hesitation, the hotel management agreed.

During our discussions on the specifics of this new menu, the head chef reassured us regarding their bakery products. "We only serve bread baked off at the hotel," he explained with visible pride. We found his choice of the word "baked off" slightly curious, and upon asking where the bread was actually baked from scratch, we discovered that this took place in Poland. Clearly, there was still room for improvement.

The irony was hard to miss. Only a few years earlier, the legendary British TV chef Keith Floyd had filmed his *Floyd's Fjord Fiesta* at the very same hotel. Standing by the Sognefjord, Floyd had famously declared the local langoustines (*Nephrops norvegicus)* to be the finest ingredients he had ever encountered in his globetrotting career. It seemed a profound paradox that a hotel situated in "nature's own pantry", once hailed by world-class chefs, had succumbed to the logistical convenience of importing frozen bake-off dough from Central Europe.

By the end of our three-year collaboration, however, the hotel had exceeded our expectations. They didn't just meet our requirements; they embraced them as a core part of their identity. Today, the Kviknes Hotel proudly presents its culinary philosophy on its website: "At the Kviknes Hotel, we place great emphasis on the culinary experience. Whenever possible, we use local ingredients—the Sognefjord is, after all, an excellent pantry!" While we may not have succeeded in shifting the entirety of Norwegian tourism onto a sustainable path, we did, at the very least, manage to transform the food philosophy at the historic Kviknes Hotel.

Scenarios for Sustainable Tourism

The work on developing scenarios for sustainable tourism took the form of creating a PC-based game—Sustainable Tourism Norway (SUSNOR)—based on Powersim software. The idea was to invite users to play their way toward scenarios and to use this activity as a valuable supplement to our own "expert-constructed" scenarios. The starting point for the scenario model (and the game) was the following three objectives from the 2007 tourism strategy:

- Increase international tourist arrivals by 1 million (as specified in our scenario model to occur by 2025).
- Make Norway a carbon-neutral destination by 2030 (the scenario model converted this to a targeted 30% reduction in CO_2 emissions from tourism compared to 2005).
- Maintain or preferably increase post-tax profits in the Norwegian tourism industry.

We placed great emphasis on collecting and systematizing relevant data in order to make the PC game as realistic as possible. This enabled us to conduct trend analyses and projections of the most likely scenarios[37]—which, several years later, proved to be accurate.

To understand how the tourism industry will develop in response to climate change, we used scenarios from the UN Intergovernmental Panel on Climate Change

[37] Peeters et al. (2011).

for global socioeconomic development—the Special Report on Emissions Scenarios (SRES)—as well as the emission scenarios themselves. In dialogue with stakeholders involved in the project, we then identified existing and prospective national climate measures within the areas of transport, technology, taxes and fees, and marketing (see Table 4.4).

Table 4.4 Norwegian climate measures that can be selected and scaled in the game SUSNOR[38]

Areas of action	Measures
Investment in the transport system	• Invest in airport capacity (0–2000 × 10^6 euros/year) • Invest in public transport (bus, train) capacity (0–2000 × 10^6 euros/year) • Invest in the development of high-speed rail (0–1000 × 10^6 euros/year) • Establish fully integrated public transport routes (yes/no) • Increase frequency of public transport (factors 1–10) • Increase user-friendliness in public transport (factors 1–10)
Technological measures for energy efficiency	• Renew domestic aircraft fleet (reduce existing service life by 0–10 years) • Renew international aircraft fleet (reduce existing service life by 0–10 years) • Renew domestic public transport fleet (reduce existing service life by 0–10 years) • Renew international public transport fleet (reduce existing service life by 0–10 years) • Increase the share of more energy-efficient turboprop aircraft in domestic operations (0–100%) • Increase efforts on energy-saving measures in domestic public transport (0–30%) • Increase efforts on energy-saving measures in accommodation services (0–30%) • Increase efforts on energy-saving measures in activity services (0–30%)
Taxes and fees	• Unilateral Norwegian carbon tax (0–1 euro/kg) • Tax or subsidy on domestic air travel (−50 to + 200%) • Tax or subsidy on domestic car travel (−50 to + 200%) • Tax or subsidy on domestic public transport (−50 to +200%) • Tax or subsidy on international air travel (−50 to + 200%) • Tax or subsidy on international public transport (−50 to +200%)
Marketing Norway as a tourist destination	• Prioritization of marketing (domestic, short trips, medium-length trips, long trips, all of Europe, equal emphasis on all markets) • National marketing budget (0–300 × 10^6 euros/year)

[38] Hille et al. (2011).

Our model estimated that if Norway continues its current path—a business-as-usual scenario—CO_2 emissions from tourism in Norway could increase by as much as 66% by 2050, compared to 2005 levels. At the same time, the model projected a substantial increase in the number of tourists, by 230% to 12.9 million arrivals. We also found that regardless of which global development and climate policy scenarios we used as a basis, the model consistently showed that the number of international tourists would at least double, profitability would increase by at least 50%, and emissions would not decrease.

To investigate how we can achieve a more desirable development—namely, to combine an ambitious climate policy with increased profitability in the tourism industry—we conducted several alternative simulations, all combined with an ambitious Norwegian climate policy targeting tourism. The simulations produced two results:

- The combination of *low* population growth and *high* global economic growth results in a *small* reduction in CO_2 emissions (6%) and a *large* increase in both the number of foreign tourists and the profitability of the industry (nearly 100%).
- The combination of *high* population growth and *low* global economic growth leads to a *greater* (though still not very high) reduction in CO_2 emissions (27%), while both profitability (60%) and the number of foreign tourists (50%) still increase significantly.

Overall, our findings emphasize that global trends in income levels and population size significantly affect the development of sustainable tourism in Norway. Furthermore, the model suggests that even highly ambitious international and national climate measures have relatively minor consequences for the profitability of the Norwegian tourism industry. At the same time, the model shows that regardless of how stringent the national climate measures implemented in our model were, it was not possible to achieve the climate target. The model simulation results indicate that much stronger policy instruments than those included in the model are needed to reach climate goals for tourism development aligned with international climate targets. The model demonstrated a clear conflict between the current trajectory of Norwegian tourism and Norway's official ambition to achieve a carbon-neutral tourism sector by 2030.

"I Can't Get the Emissions Down!": Gaming the Sustainability Gap

At the final Balestrand Summit in 2010, we introduced the participants to our bespoke PC game, SUSNOR, and invited them to play. During the coffee break before the session, the atmosphere was one of indulgent skepticism. I noticed many attendees smiling and shaking their heads, mumbling comments like "This is typical researcher stuff" and "This will never work in practice".

After a thorough briefing, we set aside one hour for everyone to test the simulation on their own laptops. One hour later, the atmosphere had shifted entirely. It was nearly impossible to tear people away from their screens. Most continued to play long into their scheduled break.

Jan Heggheim, a seasoned professional who was then leading tourism development for the former county of Sogn og Fjordane, provided the most telling feedback. Initially enthusiastic, he began to show signs of genuine frustration as he grappled with the simulation's logic. "No matter which climate measures I select," he remarked, "the number of foreign tourists just keeps increasing." Then he added, with a mix of bafflement and irritation, "And no matter what I do, I can't reach the emissions reduction target! I thought it was just a matter of introducing sufficiently high taxes!"

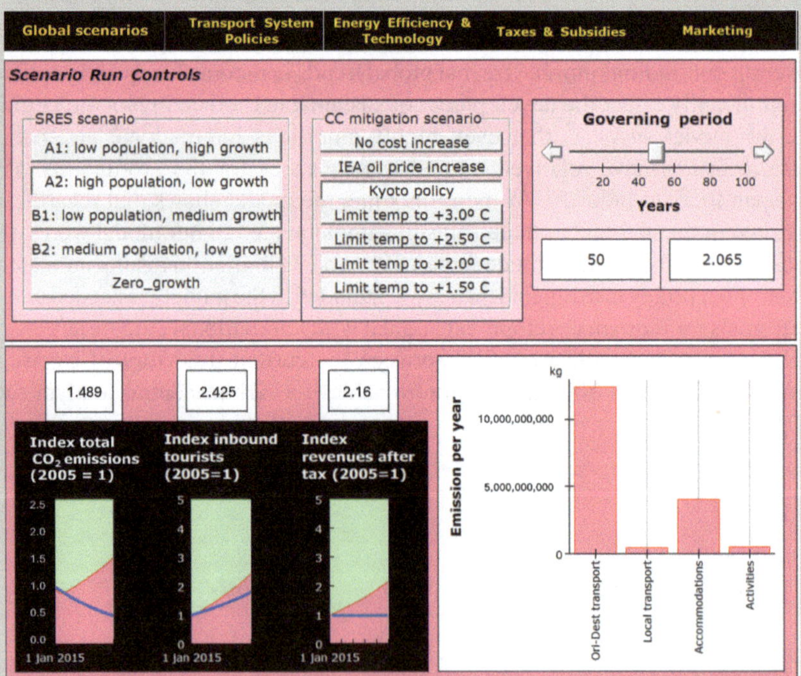

The opening screen of the PC game. The blue lines in the figures for the three indices "total CO_2 emissions," "inbound tourists," and "revenues after tax" indicate the game's objectives.

Nine years after the launch of SUSNOR at the Balestrand Summit, I conducted a workshop at a similar national tourism conference in the mountain village and tourist gem of Lom, organized in part by the Norwegian Mountain Centre. While preparing for the workshop, I used the PC game to compare actual development as of 2019 with a business-as-usual (BAU) scenario for 2025. Table 4.5 highlights three interesting aspects. *First*, official statistics show that the growth in Norwegian tourism, measured by the number of foreign visitors, far exceeds the government's target set in 2007. The increase aligns more closely with SUSNOR's BAU scenario, which assumes maximum effort to promote tourism growth and minimal effort to reduce CO_2 emissions. *Second*, the actual economic development of the tourism sector is better than expected. *Third*, the actual trend in greenhouse gas emissions unfortunately also aligns with SUSNOR's BAU projection and therefore remains very far from the government's target.

Food and Tourism: A Success Story

As already suggested in the discussion of the Balestrand Summit and how the Kviknes Hotel shifted its food profile in a more sustainable direction, I argue that Norwegian food history over the last few decades represents a genuine sustainability

Table 4.5 Comparison of scenarios for development from 2005 to 2025 created using the SUSNOR game and actual development from 2005 to 2019

Timeline	Number of foreign tourists to Norway	Total net profit in the tourism industry after tax	CO_2 emissions from Norwegian tourism
2005 actual situation (initial values entered in SUSNOR)	3.9 million	123 billion NOK	9.78 million t
2019 actual situation (calculated from official statistics)	+48%	+20%	+15–30%
2025 BAU scenario (projection from SUSNOR)	+50%	+59%	+27%
2025 National Targets (entered in SUSNOR)	+20%	Maintain 2005 level	–30%

success story in Norwegian tourism. The Winter Olympics in Lillehammer marked the beginning of this positive development. Local food and Norwegian culinary traditions, combined with a strengthening of the culinary arts, are key elements of this shift. The individual who has probably contributed the most to this change—chef Arne Brimi—also participated in efforts to give the Lillehammer Olympics an environmental profile.[39] He, along with several others who followed, brought about a significant increase in expertise and pride within Norway's culinary profession. This is illustrated by the fact that Norway has won more medals than any other country since the World Championship in Culinary Arts (Bocuse d'Or) began in 1987, with the first of a total of 11 Norwegian medals awarded in 1991.

As Norwegian culinary arts were being strengthened, a wide range of development and restructuring initiatives within agriculture led to the proliferation of new and more diverse food products. This development can be illustrated through the history of Norwegian cheese. A few decades ago, Norwegian agriculture primarily produced two (many would argue unremarkable) types of cheese—brown and yellow. Today, the situation resembles that of France, culminating in Norway's first victory at the "World Cheese Awards" in 2016 with the blue cheese Kraftkar, produced by the company Tingvollost, located in the eco-municipality of Tingvoll (see the chapter on municipal environmental policy). According to the organization Norsk Gardsost, there were over 400 varieties of cheese in Norway in 2025, produced by just over 150 producers. The number of cheeses has more than doubled since Norsk Gardsost was established in 2000. These new local cheeses appear widely in the offerings of Norwegian tourism enterprises.

From Food Shame to Food Pride: A Journey Through the Norwegian Pantry

In the summer of 1957, my American mother, Carol Claybaugh, embarked on a grand tour of Norway. She was accompanied by her future mother-in-law, Lilly Weiser Aall—then a curator at the Norwegian Museum of Cultural History—and my father, Cato Aall. Their vessel for this journey was a German-bought 1936 Opel Olympia convertible. As an American citizen, my mother had been able to import the car without the restrictions that would limit private car purchases for Norwegians for another four years.

Lilly and Cato were eager to showcase the rugged beauty of the Norwegian tourist landscape. My mother documented the journey—from Oslo through Fagernes to Lærdal and Sogndal—in a long letter home to her parents. In a twist of fate, she was describing the very region where, 33 years later, I would begin my career as a researcher at WNRI.

[39] Brimi (1993).

In the travel letter, now part of the collection at the Norwegian Travel Museum in Balestrand, she reflected on the culinary state of the nation:

> To one who is used to the charming guesthouses of central Europe, the cafés here in the North are a great disappointment. There is no tradition of coffeehouses or taverns because there were no villages in former times. The farms were scattered about the land just as they have always been in America too, and each farm had its own bakery, brewery, smithy, and guest house. The traveler—if there were travelers in those days—must have turned in at the nearest farm for refreshment and lodging.

Her observation captured the exact seed of what would eventually become a significant upgrade in Norwegian tourism: the rediscovery and professionalization of Norwegian farm food. This movement sought to unearth the knowledge and traditions dormant in the scattered farmsteads she described. In 1998, an organization with that very name (Norsk Gardsmat) was established. Following a series of mergers with rural tourism and inland fishing associations between 2006 and 2009, it formed HANEN (a nationwide trade association for rural tourism and local food, named after the iconic farmyard rooster). Today, this umbrella organization serves as the backbone for local quality products, rural tourism, and inland fishing, finally providing the charming and authentic experiences my mother found missing decades ago.

Our entry point into research on the topic of food was through studies in the late 1990s on the societal conditions for increased production and marketing of organic agricultural products, as well as systems to strengthen cooperation in marketing and sales among small-scale food producers. One finding documented a distinctly Norwegian and deep-seated skepticism toward organic farming among major agricultural stakeholders. In a study aimed at explaining why organic food product sales in Norway are much lower than in our neighboring country Sweden, we found that while Norwegian dairies aim to produce organic milk solely based on current market demand, the Swedish dairy company Arla developed its production and marketing strategy based on the idea that, in the long term, all milk production would have to comply with organic production criteria.[40]

[40] Brendehaug and Groven (2000).

When Western Norway Research Institute Led the News on Italian TV

In the early 2000s, WNRI became involved in several projects under the auspices of Euromontana, a multi-stakeholder network dedicated to sustainable development in mountain regions. As part of an EU initiative to develop new regional policies, we participated in the European research project Euromountains.net. Our goal was to analyze the role of regional authorities in promoting mountain quality products. One of our case studies focused on the Vik Dairy, located in my neighboring municipality. It is the only remaining dairy in Norway still producing the traditional cheese Gamalost.

The name literally translates to *Ancient Cheese*—not a reference to its aging process but to its deep historical roots reaching back to the Viking Age. It is a unique, fat-free mold-ripened cheese found almost exclusively in Norway.[41]

In 2004, Karl Georg Høyer, then the head of environmental research at our institute, attended an international conference in Turin organized by Euromontana. The organizers had gone to great lengths to secure an Italian minister to open the proceedings. However, as soon as the minister finished his address, with the TV cameras still rolling, Karl Georg unexpectedly stole the spotlight. Presenting our findings on Gamalost, he concluded with a remark perfectly calibrated for the Italian media:

> The defining characteristics of local food are that it often looks ugly, smells bad—and is good for sex.

Predictably, the journalists were enthralled. The minister was instantly forgotten, and Karl Georg became the lead story on the evening news.

The scientific (or perhaps semi-scientific) point behind this bold claim was the traditional folk belief in the cheese's aphrodisiac properties—a notion that has since earned it the nickname "Norway's Viking Viagra". Interestingly, a grain of scientific truth even supports the legend. A Norwegian study from 2014 indicated a correlation between Gamalost consumption and lower systolic blood pressure. With a degree of generosity, that effect could support the traditional claims, at least for older men.[42]

Throughout the 2000s, the topic of food became an important part of our research on sustainable tourism, illustrated by the metaphor of the three major Bs—Biff, Bil, Bolig (Beef, Car, Housing)—as first formulated in connection with our research on local environmental protection.[43] The three Bs represent the three areas of consumption that contribute most significantly to the climate footprint of our consumption: what we eat, how and how much we travel, and how we live. This principle also applies to leisure and tourism consumption. Based on an international literature review and our own local analyses, we developed a list of tips for making food

---

[41] Weinback et al. (2006).

[42] Nilsen et al. (2014).

[43] Aall et al. (2001).

services in tourism more climate-friendly. In retrospect, it has become evident that the Norwegian hospitality industry has followed many of these recommendations.

Today, "local food" has almost become a hallmark of food service in Norwegian tourism, even though the word "local" has often been distorted to mean merely "from a known location", rather than the more precise "sourced within a short distance". That is, as long as the origin of the raw material appears on the label—regardless of how far it lies from the place of service—the food can be presented as "local." While vegetarian menus were previously reserved for speciality restaurants in Norway, they have now become a natural part of the menu at most restaurants. However, organic ingredients still struggle to gain the same foothold in the food service industry as local food has achieved. The proportion of organic land relative to total cultivated agricultural land in Norway has remained at roughly the same level over the past thirty years (below 5%). Our studies also indicate that tourism businesses' use of food products labeled as locally produced, organic, experiential, traditional, or regionally produced is primarily marketing-driven and not mainly part of a climate or environmental strategy.[44] Based on a combination of literature studies and our own empirical research, we developed a list of recommendations on how to make food service in the tourism industry more climate-friendly (see Table 4.6). The scientific article[45] based on these studies is among the most cited from our research on sustainable tourism, with over 500 citations registered on Google Scholar.

Climate Change Adaptation on the Agenda in Norwegian Tourism

In 2017, the government presented its third white paper on tourism. Considerations regarding climate change and the need to adapt to it were only briefly mentioned (a quarter of a page out of 103 pages). The report points out that nature-based tourism and winter tourism may face challenges but concludes (p. 30)[46]: "[I]f these factors are taken into account, Norway will continue to be an attractive destination for tourists in the future."

Five years later, in 2021, Innovation Norway, commissioned by the government, presented a new national tourism strategy. However, this document contains even less information about climate change and the need for climate change adaptation in the tourism sector. The only reference to how climate change may affect Norwegian tourism is the following brief statement (p. 42)[47]:

> Climate change will also introduce new guidelines for land use in the future, and the debate on restrictions on tourism is already underway (regarding cabin construction, the extent of infrastructure development, installations in nature, and more).

---

[44] Hille et al. (2009).

[45] Gössling et al. (2011).

[46] Næringsdepartementet (2017).

[47] Innovasjon Norge (2021).

Table 4.6 Seven recommendations on how to make food services in the tourism industry more climate-friendly

What changes?	How?	Climate considerations
Dishes	Less meat, more vegetables	Greenhouse-grown vegetables remain a source of concern
	Follow the seasons	Avoid emissions associated with storage, processing, and long-distance transport
Raw materials	More locally sourced food	Transport should be energy efficient and the distance be kept short
	More organic food	This recommendation has positive environmental effects beyond reduced greenhouse gas emissions
Preparation	Reduce food waste	This is the most reliable climate advice
	Improve energy efficiency in the kitchen	This is also a reliable climate recommendation
	Prepare food from scratch	Avoid emissions associated with storage and processing, provided that locally sourced food is chosen and the kitchen is energy efficient with minimal waste

The first major Norwegian research project on how climate change may affect Norwegian tourism and relevant measures to address these changes began in 2018. Called Impacts of Climate Change on Norwegian Nature-Based Tourism (Clim-Tour), the project was led by WNRI and conducted in collaboration with Nordland Research Institute, NORCE, the Norwegian Meteorological Institute, the Department of Biosciences at the University of Oslo, and the consultancy Miljøfaglig Utredning. Many tourism businesses also participated in the project. The project revealed two main findings. The *good* news was that small-scale, nature-based experience industries have a strong capacity for adaptation. Guiding companies can find new locations to take tourists if one area loses its appeal due to climate change. The *bad* news was that winter tourism—especially infrastructure-based forms of winter tourism—will face major challenges in the future, even in Norway, a country known for its winter tourism.

We mapped all Norwegian ski resorts to assess whether they would be able to meet the so-called 100-day rule—a rule that Swiss banks (but not Norwegian ones) use when processing loan applications for the establishment of new ski resorts. The rule requires loan applicants to document that, in the future, under a changing climate, they will be able to maintain operations for a minimum of 100 days per year with natural snow, supplemented as necessary (in practice, almost always) by artificial snow production. Based on this rule, we concluded that only those ski resorts in Norway located at high altitudes or near a glacier (as in Sogn, where I am fortunate to live) could meet this requirement. Lower-lying resorts, especially those located along the coast of Western Norway, would not be able to satisfy the 100-day rule during the 2050–2080 period.[48] We were unable to include the costs of artificial snow production in the national analysis; however, we did account for these costs in a follow-up in-depth study of seven ski resorts in the areas around Sogn (Harpefossen, Stryn, Jølster, Sogn, Sogndal, Myrkdalen, and Hemsedal). This study showed that three of these resorts—the lowest-lying ones—will likely have to cease operations as early as around 2030 due to increased costs associated with artificial snow production.[49] It is reasonable to assume that this will apply to several other resorts included in our national survey. The sharp increase in electricity prices that Norway (and the rest of Europe) has experienced following Russia's invasion of Ukraine in the winter of 2022 has exacerbated this problem.

The ClimTour project has documented that ski tourism, including activities outside established resorts, may be strongly affected by climate change. This could also affect the well-being and quality of life of residents in these areas. Norwegians are, as is well known, said to be born with skis on their feet. The project developed a new service linked to the official Norwegian website for presenting climate projections, www.klimaservicesenter.no. The service received the prosaic name "Snow, Sleet, and Rain in the Future" and presents projections of precipitation and snow conditions for 11 key tourism destinations in Norway: the Lysefjord area, Møre, Helgeland, Lofoten, Trysil and Hafjell, the Røros area, Jotunheimen, Folgefonna and Hardanger, Inner Northwest Norway, the Lyngen area, and the Finnmarksvidda plateau. The outlook for Lofoten appears particularly bleak (see Fig. 4.3). Lofoten could become "black" in winter within the next 50 years, with almost no snow during the dark season, which will affect the rapidly growing tourism industry in that area and likely also the well-being of its residents.

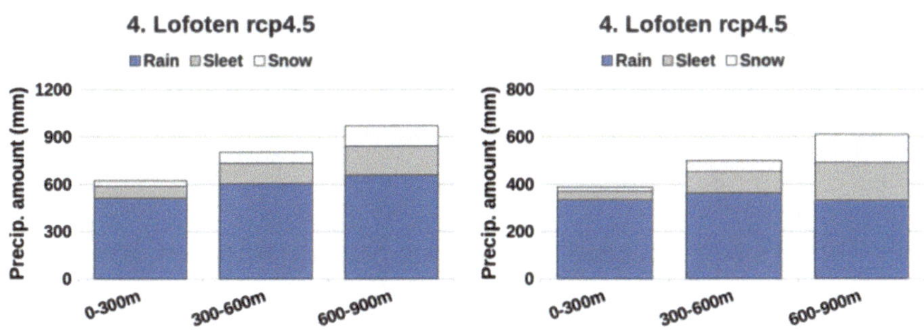

Fig. 4.3 Future prospects for Lofoten regarding snow and skiing conditions[50]

[48] Gildestad et al. (2017).

[49] Dannevig et al. (2020).

[50] Norwegian Centre for Climate Services (n.d.).

One Person’s Loss, Another’s Gain: The Rise of the Coolcation

In the summer of 2025, as relentless heatwaves and wildfires once again devastated Southern Europe, the USA, and Canada—all vital markets for our tourism industry—I was interviewed by the media regarding a burgeoning travel phenomenon: the *coolcation*.

Increasingly, tourists are choosing Norway specifically to escape extreme heat, finding that what we Norwegians traditionally dismiss as *bad summer weather*—rain and overcast skies—is precisely what they are seeking.

To satisfy my own curiosity about how fast this trend was moving, I recently did a little experiment. I set up an automated search to scan the most popular travel guides and summer itineraries that Google serves to prospective tourists looking up Norway. What I found was startling. Back in 2023, the word coolcation (or any sales pitch about escaping the heat) was virtually non-existent—appearing in less than two percent of these guides. Fast forward just two years to 2025, and the concept had absolutely exploded, taking center stage in nearly half of all the top travel itineraries I analyzed. It has gone from an obscure industry buzzword to a dominant travel hook almost overnight.

There is, however, a profound moral discomfort in this trend. Innovation Norway, our state-funded tourism agency, has leaned heavily into this shift with campaigns urging travelers to “Escape the heat on a coolcation to Norway”. While economically lucrative, using global climate distress as a marketing hook feels deeply unsettling to me.

The data from 2025 highlights the scale of this shift. Statistics Norway reported the highest number of overnight stays ever recorded for the month of June. Simultaneously, our Meteorological Institute described the month as a typical and normal Norwegian June—a mix of sun and rain with moderate temperatures. Meanwhile, across the rest of Europe, it was recorded as one of the warmest Junes in history. What was once considered a tourist’s disappointment has suddenly become a premium commodity.

Why Did We Not Succeed in Creating Sustainable Tourism in Norway?

How Bad Is It for Norwegian Tourism?

It is extremely challenging to establish a greenhouse gas inventory for the tourism sector. This task can be approached in a variety of ways. First, one must decide whether to measure production or consumption. Then, it is necessary to determine which activities to include, particularly whether travel into and out of Norway

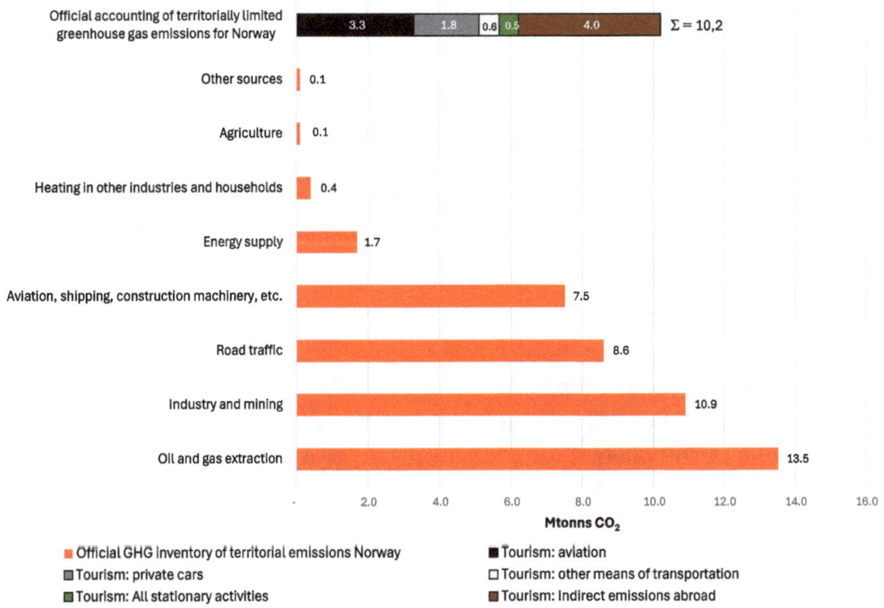

Fig. 4.4 Emissions accounting for the Norwegian tourism industry compared with official CO_2 inventory of Norway, 2019 (Mt CO_2)

should count. Data collection also presents difficulties due to limited availability. For example, no national statistics exist on transport work performed by international flights; only the number of flight movements is reported. This makes it difficult to calculate the actual amount of transport carried out by the largest source of greenhouse gas emissions from tourism.

The most recent calculation conducted by WNRI focuses on the Norwegian tourism industry. It includes activities both domestically and abroad, as well as both direct and indirect emissions.[51] The calculation shows that in 2019, direct tourism-related emissions taking place inland accounted for at least 14% of official Norwegian CO_2 emissions. If we also consider indirect emissions taking place abroad related to Norwegian tourism, this places emissions from the Norwegian tourism industry in 2019 (at a minimum) on par with the second largest source—industry and mining—in the official Norwegian CO_2 emission inventory (see Fig. 4.4). The same calculation shows that greenhouse gas emissions from the Norwegian tourism industry increased by 46% between 2007 and 2019, while emissions from Norwegian business activities decreased by 3% over the same period. If this trend continues without new measures targeting aviation emissions, tourism will become the sector of the Norwegian economy responsible for the highest Norwegian greenhouse gas emissions by 2030.

[51] Sun et al. (2022).

I have used the phrase "at least" for two reasons. First, emissions from aviation include only Norwegian airlines (that is, not the travel of Norwegians and foreigners within, to, and from Norway using foreign airlines). Second, the radiative forcing (RF) factor—which accounts for the additional climate warming effects of non-CO_2 emissions at higher altitudes, such as contrails and nitrogen oxides—has not been included. The combined effect of adjusting for these two reservations could result in aviation emission figures that are 2.5–3 times higher. Consequently, total emissions from the Norwegian tourism industry would equal the largest source of emissions in the official Norwegian emissions inventory: emissions from Norwegian oil and gas production.

Climate Transition: Washing or Greenwashing?

At regular intervals, Visit Svalbard organizes a conference titled Friske Fraspark (Fresh Impulses). In 2018, I was invited to give a 15-min presentation on how Norwegian tourist destinations could improve their climate performance. Given the carbon footprint and the time required for such a journey—the trip from the mainland would have consumed three full working days—I requested to deliver my talk digitally.

The response was a firm "no". The organizers insisted on my physical presence.

As the date approached, I found myself in a paradoxical position: to deliver a lecture on reducing emissions, I was being forced to generate a significant amount of them. Ultimately, I resorted to a white lie, claiming illness to ensure I could present via a digital link. The title of my presentation remained unchanged, though it felt more relevant than ever: "How to Talk to Tourists About Climate Without Greenwashing".

Norway Has Reached Its Goal—According to Some

Many believe that Norway has already achieved the transition to sustainable tourism, or at least stands very close. Both the government and the largest industry organization in tourism (NHO Reiseliv) point out that the number of tourists in Norway is increasing—both domestic and international—that the economy is recovering in the Norwegian tourism sector after the catastrophic years of the COVID-19 pandemic, and that electric mobility is on the rise (including passenger cars and ferries, and soon, according to the most technology-optimistic, even electric aircraft). They acknowledge only two caveats: the problems and conflicts surrounding cruise tourism (which, at the same time, continues to grow), and a nascent concern that Norway, too, may experience some of the issues of overtourism that increasingly affect destinations across Europe.

The Norwegian variant of overtourism is, characteristically, not primarily found in cities but rather in rural areas. Narrow roads crowded with slow-moving camper vans and tourist buses, full ferries (many of which have been made free to facilitate living in rural Norway), and parking lots at Instagram hotspots that become overcrowded—all create traffic chaos and frustration among residents who use the same roads and ferries in their daily lives. Overtourism has also become an issue for the most visited hiking trails, such as Preikestolen (300,000 visitors), Trolltunga (80,000–120,000 visitors), and Besseggen (60,000–70,000 people).

Short-Distance Tourism Pays Off: The Case of Gjendesheim

During the 2010s, significant resistance emerged within parts of the tourism industry against Innovation Norway's marketing campaigns targeting overseas markets. In 2018, the Norwegian Trekking Association (*Den Norske Turistforening* [DNT]) made a landmark decision: they ended their collaboration with Innovation Norway on marketing aimed at long-haul travelers, particularly from Asia and the United States. Several private operators specializing in nature-based tourism followed suit, withdrawing from joint campaigns. Their grievance was clear: they felt that the focus on overseas mass tourism undermined their efforts to attract more affluent, environmentally conscious European and domestic visitors.

Marius Haugaløkken, the manager of Gjendesheim—the DNT's second most visited staffed lodge—became a vocal critic of this volume-driven policy. In the summer of 2021, while my son was working a summer job at the lodge, he recounted a telling story from Marius that illustrated the success of this defiance.

Despite the ongoing COVID-19 pandemic, Gjendesheim performed exceptionally well financially that summer, catering to Norwegian and Nordic tourists. This was no accident. Years earlier, they had deliberately shifted their marketing strategy to focus on the regional market. Previously, skeptics had shaken their heads, claiming, "If we stop marketing Norway to the Chinese, they will simply go to Sweden instead." The irony was complete when Innovation Norway, intrigued by Gjendesheim's resilience and profitability, arrived on a study tour. They wanted to learn how the lodge had succeeded by doing the exact opposite of the national agency's own overseas strategy.

According to Others, We Have Not Reached the Goal—Why Not?

Some—including myself—believe that Norwegian tourism has not yet succeeded in transitioning to sustainability. The main argument for this position is that greenhouse gas emissions continue to rise, in contrast to the trend observed in many other Norwegian industries. This increase is linked primarily to dramatic growth in both air and cruise tourism, a growth that more than offsets the gains in energy efficiency achieved in aviation and maritime technology. The rise in emissions also continues despite Norway’s global leadership in the transition to electric mobility, both in passenger vehicles and ferry transport.

What accounts for this negative trend in environmental sustainability within Norwegian tourism? As with the development of the SUSNOR scenario model for the evolution of Norwegian tourism, it is useful to distinguish between international and domestic drivers.

After the project Sustainable Destination Norway 2025 concluded around 2010, I gave several presentations in various contexts, outlining strategies and measures to make tourism more sustainable. In the summer of 2025, at one of WNRI’s events during Arendalsuka, my colleagues presented recommendations in connection with the conclusion of our second major overarching project on sustainable tourism, A Resilient Low-Carbon, High-Yield Tourism Model for Norway (TourRisk). Our recommendations regarding the need for policy change are depressingly similar. One could interpret this as us researchers failing to renew ourselves. The other, and unfortunately more reasonable, interpretation is that Norwegian tourism policy has failed to renew itself. Therefore, repeating our recommendations from 15 years earlier is entirely appropriate.

As the SUSNOR model simulations identified, the development of international supply and demand for air travel has significant implications for sustainability efforts in Norway. Internationally, this trend has increased drastically during the period since work began to transition Norway’s tourism sector toward sustainability. Global CO_2 emissions from air travel increased by 580% from 1960 to 2018—exactly twice the percentage increase of total global emissions.[52] This driver is further reinforced by the lack of international policies addressing global greenhouse gas emissions from air travel.[53]

[52] Lee et al. (2021).

[53] Gössling and Humpe (2023).

Hope Shattered: On the COVID-19 Pandemic as a "Perfect Storm"

The expression *a perfect storm* originally described extraordinary and coinciding weather phenomena that were particularly damaging. Gradually, it has acquired a metaphorical and, in fact, opposite meaning, characterizing critical events that can create opportunities to reverse a situation or development in a more positive direction. Such an expectation was also expressed in the report *One Planet Vision for a Responsible Recovery of the Tourism Sector*, published by the UN's tourism body during the COVID-19 pandemic.[54] The pandemic had massive consequences for the tourism industry. Global flight departures in April 2020 were 75% lower than in 2019 before the outbreak—a situation that also led the International Civil Aviation Organization (ICAO) to weaken its own rules for CO_2 compensation.

A recently completed doctoral dissertation that I had the pleasure of supervising showed that Norwegian government authorities implemented two very different types of strategies to address the dramatic challenges facing the tourism industry as a result of the pandemic: support for survival (what the dissertation termed a "bounce-back resilience strategy") and support for restructuring (a "bounce-forward resilience strategy").[55] The author's investigations into how Norwegian authorities responded to the pandemic demonstrate that the survival strategy received by far the largest share of funding. The support consisted of NOK 20–25 billion in compensation schemes for businesses, wage support schemes, expanded furlough opportunities, and a dedicated scheme for the aviation industry. This included NOK 6 billion in loan guarantees for Norwegian airlines, direct purchase of flight routes to maintain a minimum level of domestic service, and access to various other compensation schemes (such as temporary exemption from the air passenger tax and reductions in aviation fees). The development strategy accounted for about 5% of the survival strategy in terms of support amounts and was channeled through Innovation Norway in the form of extraordinary marketing support to stimulate domestic tourism, liquidity loans and grants to small and medium-sized tourism enterprises, and dedicated so-called restructuring funds. Only a small portion of Innovation Norway's funds carried requirements linked to the goal of sustainable tourism, and then only in the form of requirements to use the restructuring funds to create more environmentally friendly ("greener") operations, new (more environmentally friendly) products (such as increased use of local or organic ingredients in food services), and support for developing a generally more sustainable business model.

In 2025, three years after the lifting of COVID-19 restrictions in Norway and most of the world, the volume of leisure air travel was back on track and even surpassing previous levels. According to the IATA, total global air traffic in 2024—the first full year without COVID-19 restrictions on air travel—measured by passenger numbers, was 6.8% higher than in 2019, and 3% higher in Europe, while the situation in

[54] UNWTO (2020).

[55] Aamot (2025).

Norway was 5% lower according to Avinor. The COVID-19 pandemic has clearly not contributed to the sustainability transition in tourism that some of us may have hoped would emerge.

Norwegian Technological Optimism

Measures implemented specifically to reduce greenhouse gas emissions from tourism in Norway have primarily taken the form of technological initiatives that have led to a slight decrease in relative emissions (per tourist, per kilometer traveled, etc.). Meanwhile, total emissions have nonetheless increased. As previously noted, Norway is a world leader in the transition to electric mobility, a development that has understandably influenced the tourism sector. So far, however, this has had only a marginal effect on emissions from leisure travel by private car and ferry within the country. There has been renewed discussion about shore-power and cruise ships, but here too, the impact of measures has so far been marginal, as few initiatives have been implemented. Some discussion has also occurred about electric aircraft, but this has had no effect whatsoever to date; on the contrary, emissions from aviation have increased dramatically.

At a presentation during the national conference on cruise tourism in Norway in 2022, I discussed a news article published by the Norwegian Broadcasting Corporation (NRK) regarding proposed hydrogen-powered cruise ships. This coverage perfectly illustrate what I consider a naive technological optimism characterizing dominant actors in tourism policy, thereby blocking more effective measures. The NRK report highlighted proposed zero-emission hybrid vessels, featuring interviews with major tourism stakeholders on how they believe the cruise industry's climate challenges should be addressed. My main point was that this reflected a typical "too little, too late" approach. The article promoted a "climate solution" that would not be ready until 2025 or 2026. From 2025, Norway has only five years remaining to halve its greenhouse gas emissions and thirty years until emissions must reach nearly zero, according to our national climate targets. Moreover, the coverage focused on small-scale vessels, explicitly noting that such ships have "room for far fewer passengers." The final uncertainty acknowledged in the article was that "the technology is not yet mature." In 2020, shipyards had 118 new cruise ships under construction globally, scheduled for delivery by 2027—a record high compared to previous periods. None of these ships have zero-emission solutions.[56]

The problem with technological optimism, as discussed above, is not the optimism itself, but rather how this optimism functions in public debate and (therefore also) in policymaking. It can serve as a lightning rod and dampen discussions about measures that could have an immediate and indisputable effect, such as a ban on fossil-fuelled cruise ships in Norwegian waters, or the somewhat slower-acting but nonetheless

[56] Cruise Industry News (2024).

effective instrument of taxation. These are also politically controversial measures, and headlines like the one shown below can therefore put such initiatives on hold, based on the expectation that technology will soon arrive and solve the problem.

Optimism Prevails: A Refusal to Debate

During Arendalsuka 2024—Norway's largest political festival—WNRI hosted a seminar titled "Growth in Tourism: High Revenues, but High Emissions and Strained Local Communities. What Is the Way Forward?" We facilitated a debate centered on a pressing question: since tourism growth generates income and jobs but simultaneously drives up greenhouse gas emissions and increases pressure on local infrastructure and nature, should Norway introduce stronger national environmental guidelines for the industry?

During the planning phase, we invited Kristin Krohn Devold, the CEO of NHO Reiseliv, to participate. Her emailed refusal was a testament to the prevailing technological optimism within the sector:

> Planes, trains, cars, and ferries—the modes of transport people use to access tourism products—have achieved dramatic emissions reductions. Ferries are largely electric, cars are 50 percent electric, trains are electric, and buses are transitioning to electricity in many areas. Even aviation has made significant cuts with more modern aircraft and greater use of biofuels, as well as being required to purchase emission allowances under the EU's aviation system. Cruise ships, on the other hand, which have the highest emissions, are not Norwegian companies with Norwegian employees paying taxes to Norway. So, what is meant by the claim that emissions from Norwegian tourism have not been reduced since 2007?

The answer to Devold's rhetorical question lies in the rebound effect. While individual ferries and cars have indeed become cleaner, the sheer increase in the number of trips, the distance traveled, and the continued dominance of kerosene-based aviation have effectively neutralized these efficiency gains. In the logic of the sustainability trajectories shown in Fig. 4.1, NHO Reiseliv remains focused on the intensity axis, while the climate crisis unfolds on the volume axis.

While NHO Reiseliv declined the invitation, other key players accepted. These included Fjord Norway, the industry forum Norsk Tourism, and a Member of Parliament from the Labor Party. Evidently, they found it meaningful to debate the gap between technological hope and the stubborn reality of rising total emissions.

Norway, the Oil Nation

The Norwegian oil adventure is a driver equal in power to naive technological optimism. On the one hand, the exceptionally large state revenues generated through the taxation of petroleum activities on the Norwegian continental shelf have enabled the implementation of extensive and costly technological initiatives. According to calculations by the Ministry of Finance, cited by NRK in a report on Norwegian electric vehicle policy, the Norwegian state indirectly subsidized the transition to electric vehicles by approximately NOK 640 billion between 2007 and 2025 through reduced vehicle taxes and the exemption of electric cars from value-added tax.[57] This is, of course, a very large sum, but still corresponds to no more than the annual tax revenues the Norwegian state receives from oil and gas activities in a single year.

In other words, Norway can afford to indulge in expensive technological optimism. However, an even stronger driving force for the further development of Norwegian tourism may be the expectation that the current revenue level from taxation of Norway's oil and gas industry will be maintained even after this industry—at some point in the future—either winds down or, in the worst case (from a climate perspective), simply "runs out". This expectation is reflected in the ongoing Norwegian debate on the question: "What will we live on after oil?" The assumptions underlying this debate are well described in a report prepared by the influential consultancy McKinsey.[58] The report, titled *Norway Tomorrow*, outlines how Norway can create at least 200,000 new jobs by 2030 to replace those lost in the oil and gas sector, which, according to McKinsey, can be achieved within the following ten industries: hydrogen, offshore wind, batteries, carbon capture and storage, green maritime industries, industrial software, consumer platforms, circular economy initiatives, aquaculture, and notably, tourism.

The section devoted to tourism carries the revealing title "The Time Has Come to Industrialize Norwegian Tourism." It projects that by increasing the price level of tourism products and targeting marketing efforts toward the global middle class, the sector will contribute NOK 14 billion in increased value creation and generate 17,000 new jobs by 2030, up from the current levels of approximately NOK 130 billion and 160,000 jobs. The report highlights the large and growing greenhouse gas emissions from the tourism sector, noting that transport accounts for more than 90% of associated emissions, but at the same time asserts that the climate crisis has no dampening effect on the public's travel activity. The report recommends that tourism destinations focus on positioning themselves as "greener" than their competitors, thereby "gaining a competitive advantage" (p. 78). Of course, this report does not have the same governing authority as, for example, a white paper to the Storting or a government action plan, but it nevertheless provides insight into how key stakeholders think—or could be persuaded to think—about the future development of Norwegian tourism, with or without the word "sustainability".

---

[57] Sande (2025, May 22).

[58] McKinsey (2022).

How Can a Genuine Transition to Sustainable Tourism Be Achieved?

The way we think about and describe a problem influences how we are likely to attempt to address it and, consequently, the results we can expect to achieve. At the outset, I presented a framework that has underpinned much of our tourism research, which categorizes the sustainability challenges of tourism into mobile/stationary activities and volume/intensity issues (Fig. 4.5). This framework formed the basis for our critique of 1990s work on sustainable tourism, which we argued was too narrowly focused on stationary activities at the expense of mobile ones, and on intensity issues at the expense of volume issues.

Since then, the focus of the Norwegian tourism debate has expanded to include both volume issues and the mobile aspects of tourism. Nevertheless, we see that sustainability challenges have increased in Norway. Greenhouse gas emissions have risen—not decreased—since the debate on sustainable tourism gained traction in Norway in the early 2000s.

To understand this environmentally adverse development, we need to introduce a third axis that describes how sustainability challenges are addressed, not just how the problems *manifest* themselves—as the framework in Fig. 4.5 does. There are many pairs of concepts in the literature that—each in their own way—attempt to describe choices in how to address sustainability challenges: for example, "ecological modernization" versus "degrowth," which frequently appears in the debate on the

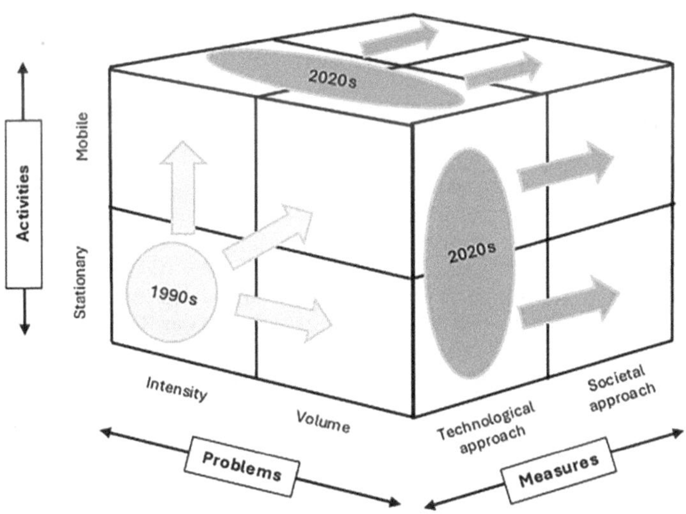

Fig. 4.5 An expanded framework for describing the sustainability challenges of tourism, which also includes the orientation of measures, indicating the dominant understanding of sustainable tourism in the 1990s and 2020s, and the need to broaden this understanding to a holistic perspective on sustainable tourism development

transition to a zero-emissions society[59]; "transition" versus "transformation," which is frequently used in the debate on energy transition[60]; and "incremental" versus "transformational change," which is often employed in the debate on climate change adaptation.[61]

Below, these dichotomies are simplified into a central distinction between *hard* strategies, which focus primarily on technological measures (referred to as "technologically approach" in Fig. 4.5), and *softer* strategies that also aim to change societal factors such as values, knowledge, organization, and consumption (referred to as "societally approach" in the figure).

Figure 4.5 illustrates a two-stage development toward sustainable tourism. The *first* stage involves broadening the understanding of which problems are addressed, from primarily stationary activities and their intensity-related issues to mobile activities and volume-related issues. The *second* stage involves expanding strategies from technology-centric measures to include those that entail more profound societal changes. Not least among these is the question of changing consumption patterns or, more accurately, reducing consumption. The latter applies to current forms of cruise tourism and (crucially) air travel tourism.

The transition from the "Best Games Ever" in 1994 to the "Coolcation" marketing of 2025 reveals a tourism policy that has matured in its rhetoric but remained stagnant in its impact. To move forward, we must stop hiding behind the shield of technological optimism and begin the challenging political conversation about limiting the total volume of high-emission mobility.

References

Aall, C., Brendehaug, E., Fløtre, T.-A., Aarethun, T., & Høyer, K. G. (2003). Næringsmessig potensiale for kvalitetsturisme [Commercial potential for quality tourism] (VF-report 4/2003). Western Norway Research Institute.

Aall, C., & Gössling, S. (2019, November 21). *NHOs klimaregnskap er to skritt frem og ett tilbake* [NHO's climate accounting is two steps forward and one step back]. Forskersonen. https://forskersonen.no/debattinnlegg-klima-meninger/nhos-klimaregnskap-er-to-skritt-frem-og-ett-tilbake/1596200

Aall, C., & Groven, K. (2004). Institutt for kvalitetsturisme. Identifisering av eit kunnskapsfelt [Institute for quality tourism] (VF-notat 3/2004). Western Norway Research Institute.

Aall, C., Holm, T. B., Jansen, M., Gobert, J., Rudolf, F., Jarris, K., & Lager, F. (2023a). Think global—act local: The challenge of producing actionable knowledge on transboundary climate risks. *Frontiers in Climate, 5*, Article 1170142. https://doi.org/10.3389/fclim.2023.1170142

Aall, C., Høyer, K. G., & Lafferty, W. M. (Eds.). (2001). *Fra miljøvern til bærekraftig utvikling i kommunene: Erfaringer med Lokal Agenda 21* [From environmental protection to sustainable development in the municipalities: Experiences with Local Agenda 21]. Gyldendal Akademisk.

[59] Xue et al. (2016).

[60] Aall et al. (2023a).

[61] Aall et al. (2023b).

Aall, C., Meyer-Habighorst, C., Gram-Hanssen, I., Korsbrekke, M. H., & Hovelsrud, G. (2023b). "I'm fixing a hole where the rain gets in, and stops my mind from wandering": Approaching sustainable climate change adaptations. *Weather, Climate, and Society, 15*(2), 349–364. https://doi.org/10.1175/WCAS-D-22-0113.1

Aall, C., & Teigland, J. (2002). Lokal Agenda 21 som nytt forvaltningsregime for å fremme et bærekraftig reiseliv? [Local Agenda 21 as a new management regime for sustainable tourism?]. In C. Aall, K. G. Høyer, & W. M. Lafferty (Eds.), *Fra miljøvern til bærekraftig utvikling i kommunene* (pp. 332–338). Gyldendal akademisk.

Aamot, D. B. (2025). Co-survival and Sustaining Tourism during Fundamental Challenges. A Case Study on the Nature-based Tourism Sector in Alta. (PhD thesis). Tromsø: UiT The Arctic University of Norway.

Avinor. (2025, August 8). *Norway as an international test arena for zero- and low-emission aviation opened with first all-electric flight in Stavanger* [Press release]. https://kommunikasjon.ntb.no/pressemelding/18612875/norge-som-internasjonal-testarena-for-null-og-lavutslippsluftfart-apnet-da-forste-flyging-med-helelektrisk-fly-ble-gjennomfort-i-stavanger

Bjørndal, O. M., & Thompson, S. (2019, November 22). *Rapporten fra NHO Reiseliv er et godt nullpunkt for videre utslippskutt* [The report from NHO Reiseliv is a good starting point for further emission cuts]. Forskersonen. https://www.forskersonen.no/debattinnlegg-klima-meninger/rapporten-fra-nho-reiseliv-er-et-godt-nullpunkt-for-videre-utslippskutt/1597125

Brendehaug, E. (2016). *Berekraftige destinasjonar. Ein studie av berekraftige destinasjonar som del av styresmaktene sin strategi for berekraftig reiseliv* [Sustainable destinations. A study of sustainable destinations as part of the government's strategy for sustainable tourism] (VF-report 4/2016). Western Norway Research Institute.

Brendehaug, E., & Groven, K. (2000). *Økologi og marknad. Ein studie av rasjonalitet i produksjon og marknadsføring av økologisk mjølk og kjøt* [Ecology and market. A study of rationality in organic milk and meat production] (VF-report 2/00). Western Norway Research Institute.

Brimi, A. (1993). *Ein smak frå norsk natur: Langs ein vassveg* [A taste of Norwegian nature: Along a waterway]. Universitetsforlaget.

Commission on Sustainable Development (CSD). (1999). *Tourism and sustainable development. Report of the Secretary-General to Commission on Sustainable Development, Seventh Session.* United Nations.

Cruise Industry News. (2024). *Gross tonnage of cruise ships in the global order book as of April 2024, by shipyard* [Graph]. Statista. Retrieved March 8, 2026, from https://www.statista.com/statistics/1111954/global-cruise-ship-order-book-by-shipyard/

Dannevig, H., Gildestad, I. M., Steiger, R., & Scott, D. (2020). Adaptive capacity of ski resorts in Western Norway to projected changes in snow conditions. *Current Issues in Tourism, 24*(22), 3206–3221. https://doi.org/10.1080/13683500.2020.1865287

Gildestad, I. M., Dannevig, H., Steiger, R., & Aall, C. (2017). *Konsekvensar av klimaendringar for norske skianlegg* [Consequences of climate change for Norwegian ski resorts] (VF-report 10/2017). Western Norway Research Institute.

Gössling, S., Aall, C., & Peeters, P. (2008). Teknologiske tiltak: Ikke nok til å redusere klimagassutslipp fra fly [Technological measures: Not enough to reduce greenhouse gas emissions from aviation]. *Samferdsel*, (5). https://samferdsel.toi.no/nr-5/teknologiske-tiltak-ikke-nok-til-a-redusere-klimagassutslipp-fra-fly-article19913-1037.html

Gössling, S., Garrod, B., Aall, C., Hille, J., & Peeters, P. (2011). Food management in tourism: Reducing tourism's carbon 'foodprint'. *Tourism Management, 32*, 534–543. https://doi.org/10.1016/j.tourman.2010.04.006

Gössling, S., Hall, M. C., Ekström, F., Engeset, A. B., & Aall, C. (2012). Transition management: A tool for implementing sustainable tourism scenarios? *Journal of Sustainable Tourism, 20*(6), 899–916. https://doi.org/10.1080/09669582.2012.699062

Gössling, S., & Humpe, A. (2023). Net-zero aviation: Time for a new business model? *Journal of Air Transport Management, 107*, Article 102353. https://doi.org/10.1016/j.jairtraman.2022.102353

Gössling, S., Humpe, A., & Sun, Y. Y. (2024). Are emissions from global air transport significantly underestimated? *Current Issues in Tourism, 28*(5), 695–708. https://doi.org/10.1080/13683500.2024.2337281

Gössling, S., & Lyle, C. (2021). Transition policies for climatically sustainable aviation. *Transport Reviews, 41*(5), 643–658. https://doi.org/10.1080/01441647.2021.1938284;

Grønt Skipsfartsprogram. (2020). *Elektrifisering av skipsfarten. Status for landstrøm i stamnetthavnene* [Electrification of shipping]. Rapport november 2020. https://grontskipsfartsprogram.no/wp-content/uploads/2020/12/Elektrifisering-av-skipsfarten-Status-for-bruk-av-landstrom-i-stamnetthavner-ZERO-Gront-Skipsfartsprogram.pdf

Hille, J., Ekström, F., Aall, C., & Brendehaug, E. (2009). *Klimamerking av mat—er det mulig?* [Climate labeling of food—Is it possible?] (VF-report 8/2009). Western Norway Research Institute.

Hille, J., Storm, H., Sataøen, H., Aall, C., & Holden, E. (2008). *Miljøbelastningen fra norsk forbruk og norsk produksjon 1987–2007* [The environmental impact of Norwegian consumption and production 1987–2007] (VF-report 2/2008). Western Norway Research Institute.

Hille, J., Vik, M., & Peeters, P. (2011). *Background for scenario making* (VF-notat 1/11). Western Norway Research Institute.

Høyer, K. G. (2000). Sustainable tourism or sustainable mobility? The Norwegian case. *Journal of Sustainable Tourism, 8*(2), 147–160. doi:https://doi.org/10.1080/09669580008667355

Høyer, K. G., & Simonsen, M. (1996). *På reisefot—men ikke til fots. Om reiseliv, transport og miljø* [On the move—But not on foot] (VF-report 7/96). Western Norway Research Institute.

Innovasjon Norge. (2021). Nasjonal reiselivsstrategi 2030. Sterke inntrykk med små avtrykk [National tourism strategy 2030]. Oslo.

Innovasjon Norge. (2024). *Merket for Bærekraftig reisemål* [The sustainable destination label]. Retrieved March 8, 2026, from https://www.visitnorway.no/planlegg-reisen/gronn-norgesferie/barekraftig-reiseliv/#steder

Johnson, S. P. (Ed.). (1993). *The Earth Summit: The United Nations Conference on Environment and Development (UNCED)*. Graham and Trotman.

Lee, D. S., Fahey, D. W., Skowron, A., Allen, M. R., Burkhardt, U., Chen, Q., Doherty, S. J., Freeman, S., Forster, P. M., Fuglestvedt, J., Gettelman, A., De León, R. R., Lim, L. L., Lund, M. T., Millar, R. J., Owen, B., Penner, J. E., Pitari, G., Prather, M. J., Sausen, R., & Wilcox, L. J. (2021). The contribution of global aviation to anthropogenic climate forcing for 2000 to 2018. *Atmospheric Environment, 244*, Article 117834. https://doi.org/10.1016/j.atmosenv.2020.117834

McKinsey. (2022). Norge i morgen. Ti mulighetsnæringer for Norge [Norway tomorrow. Ten opportunity industries]. https://www.mckinsey.com/no/norge-i-morgen

Nærings- og handelsdepartementet. (2006). *Verdifulle opplevelser. Regjeringens nasjonale reiselivsstrategi* [Valuable experiences: The government's national tourism strategy]. Oslo.

Næringsdepartementet. (2017). Opplev Norge—unikt og eventyrlig [Experience Norway—unique and adventurous] (Meld. St. 19 (2016–2017)). Ministry of Trade, Industry and Fisheries, Oslo.

NHO Reiseliv. (2019, August 28). *Cruiseturister slipper ut mest, legger igjen minst* [Cruise tourists emit the most, spend the least]. https://www.nhoreiseliv.no/vi-mener/barekraftig-reiseliv/nyhet2/2019/cruiseturister-slipper-ut-mest-legger-igjen-minst/

Nilsen, R., Pripp, A. H., Høstmark, A. T., Haug, A., & Skeie, S. (2014). Short communication: Is consumption of a cheese rich in angiotensin-converting enzyme-inhibiting peptides, such as the Norwegian cheese Gamalost, associated with reduced blood pressure? *Journal of Dairy Science, 97*(5), 2691–2693. https://doi.org/10.3168/jds.2013-7429

Innovasjon Norge. (2011). *Prinsipper for et bærekraftig reiseliv* [Principles for sustainable tourism].

Norwegian. (2025, March 3). *Norwegian to purchase ten Boeing 737-800 aircraft* [Press release]. https://media.no.norwegian.com/pressreleases/norwegian-kjoeper-ti-boeing-737-800-fly-3372846

Norwegian Centre for Climate Services. (n.d.). *Snø, sludd og regn* [Snow, sleet, and rain]. Retrieved October 25, 2023, from https://klimaservicesenter.no/kss/laer-mer/sno-sludd-regn#lofoten

NOU 2010:10. Tilpassing til eit klima i endring [Adapting to a changing climate]. Ministry of Climate and Environment.

NRK. (2017, October 23). *Shit O'hoi!* [TV Documentary]. NRK Brennpunkt. https://tv.nrk.no/serie/brennpunkt/2017/MDDP11001417

NRK. (2019, August 21). *Election 2019: Live from Flåm* [Television broadcast]. https://tv.nrk.no/serie/valg-2019-valgsending/2019/NNFA19000319

Our World in Data. (n.d.). *Number of air trips per capita* [Data set]. Retrieved October 24, 2023, from https://ourworldindata.org/grapher/air-trips-per-capita

Peeters, P., Higham, J., Kutzner, D., Cohen, S., & Gössling, S. (2016). Are technology myths stalling aviation climate policy? *Transportation Research Part D, 44*, 30–42. https://doi.org/10.1016/j.trd.2016.02.004;

Peeters, P. M., Middel, J., & Hoolhorst, A. (2005). *Fuel efficiency of commercial aircraft* (NLR-CR-2005-669). Peeters Advies.

Peeters, P., Vik, M. L., & Gössling, S. (2011). Scenarios for Sustainable Destination Norway 2025 (VF-notat 1/2011). Western Norway Research Institute.

Sabima. (n.d.). Naturkampen. Retrieved March 8, 2026, from https://naturkampen.sabima.no

Sande, B. (2025, May 22). Klimatiltak: Staten kuttet bilavgifter med 640 milliarder kroner [Climate measures: The state cut car taxes by 640 billion NOK]. NRK. https://www.nrk.no/norge/klimatiltak_-staten-kuttet-bilavgifter-med-640-milliarder-kroner-1.17441807

Simonsen, M., Walnum, H. J., & Gössling, S. (2018). Model for estimation of fuel consumption of cruise ships. *Energies*, *11*(5), 1059. https://doi.org/10.3390/en11051059

Sogn, A. (2014). *Bærekraftig reiseliv: en evaluering av pilotprosjektet Bærekraftig reisemål* [Sustainable tourism: An evaluation] [Master's thesis, Norwegian University of Life Sciences (NMBU)].

Sun, Y., Gössling, S., Hem, L. E., Iversen, N. M., Walnum, H. J., Scott, D., & Oklevik, O. (2022). Can Norway become a net-zero economy under scenarios of tourism growth? *Journal of Cleaner Production, 363*, Article 132414. https://doi.org/10.1016/j.jclepro.2022.132414

Teigland, J. (1996). Impacts on tourism from mega-events: The case of Winter Olympic Games (VF-report 13/96). Western Norway Research Institute.

Teigland, J. (1997). *Potensiell/latent etterspørsel etter fritidsreiser og sosial og kulturell endring* [Potential/latent demand for leisure travel] (VF-report 16/97). Western Norway Research Institute.

Teigland, J. (2000). *Ferielivets merkelige uforanderlighet* [The strange unchanging nature of holiday life] (VF-report 9/00). Western Norway Research Institute.

The Travel Inspector. (2022, February 18). *Den norske reiselivs-bærekraftbløffen avsløres* [The Norwegian tourism sustainability bluff revealed]. https://www.thetravelinspector.no/news/2022/2/18/den-norske-reiselivs-baerekraftbloeffen-avsloeres

United Nations World Tourism Organization (UNWTO). (2019). *Transport-related CO_2 emissions of the tourism sector—Modelling results*. https://www.e-unwto.org/doi/book/10.18111/9789284416660

UNWTO. (2020). One Planet Vision for a Responsible Recovery of the Tourism Sector. United Nations World Tourism Organization. UNWTO. (2020). *One planet vision for a responsible recovery of the tourism sector*. https://www.untourism.int/covid-19-oneplanet-responsible-recovery-initiatives

Walnum, H. J. (2011). *Energy use and CO2 emissions from cruise ships—A discussion of methodological issues* (VF-notat 2/2011). Western Norway Research Institute.

Weinback, J. E., Brendehaug, E., & Aall, C. (2006). *The role of regional authorities in the development and promotion of mountain quality products* (VF-notat 1/06). Western Norway Research Institute.

Xue, J., Walnum, H. J., Aall, C., & Næss, P. (2016). Two contrasting scenarios for a zero-emission future in a high-consumption society. *Sustainability*, *9*(1), Article 20. https://doi.org/10.3390/su9010020

Chapter 5
Norway's Climate Schism—The National Oil-Driven Policy Versus Local Consumer Action

Abstract This chapter explores the Norwegian climate paradox: the gap between ambitious national climate targets and the country's role as one of the world's largest exporters of oil and gas. The author introduces the concept of "oil-washing" to describe how national policy has prioritized technological solutions over fundamental changes in production and consumption. By tracing local climate planning from the early initiatives of the 2000s to today's professionalized frameworks, the chapter highlights the challenges of integrating consumption-based emissions into municipal governance. Drawing on experiences from the research projects HOPE and TRANSFORM, the chapter analyses the barriers to a consumption-oriented climate policy—particularly policymakers' fear that the public will resist interventions in private lifestyles. The chapter concludes that a genuine green transition requires local authorities to reclaim their 1990s role as national front-runners—outpacing both peer municipalities and the state to force shifts in national policy, challenging the idea of perpetual economic growth, and putting consumption volume on the agenda.

The Scientific and Political Context

In 1969, oil was discovered for the first time in Norway in commercially viable quantities at the Ekofisk field in the North Sea. From the early 1970s, Norway experienced remarkable growth in oil and later gas production, culminating in the record year of 2004, when Norway became the seventh largest petroleum-producing nation, with a total output of 264 million standard cubic meters (Sm^3). Since that peak year, annual production has stabilized at just under 250 million Sm^3 oil equivalents. For the period relevant to the international climate agreement—with 1990 as the reference year—Norwegian CO_2 emissions in 2024 stood 6% higher, and petroleum production was nearly twice as large as in 1990 (see Fig. 5.1).

C. Aall, *Sustainable Development*, https://doi.org/10.1007/978-3-032-35930-8_5

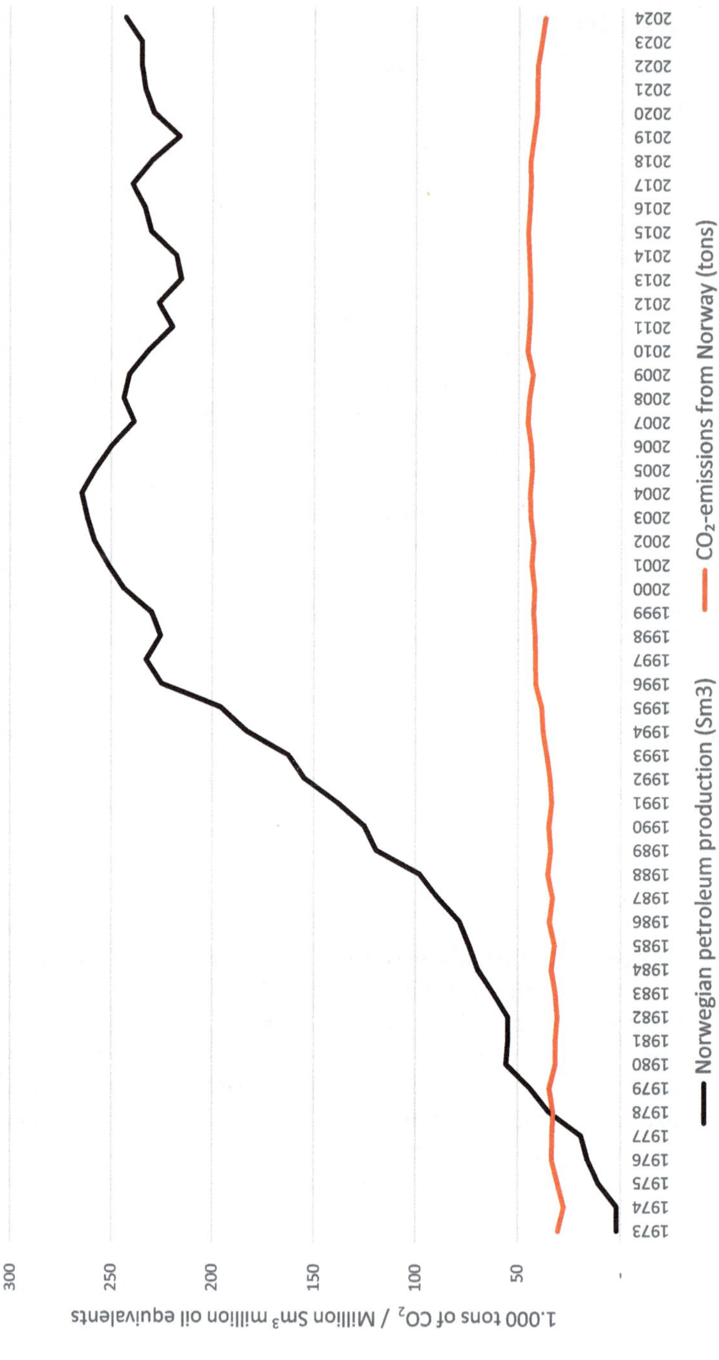

Fig. 5.1 The two faces of Norway: production of petroleum (oil, condensate, natural gas liquids, and gas) and national emissions of CO_2 from 1973 to 2024[1]

[1] Norwegian Offshore Directorate (2024) and Statistics Norway (2024)

Figure 5.1 illustrates the phenomenon that Swedish investigative journalist Lisa Röstlund refers to as the "Norway Paradox," detailed in her book of the same name, subtitled "An Investigation into the Dark Side of the Good Country."[2]

This insight is neither new nor unfamiliar to Norwegian political actors. As early as the 1990s, Norwegian authorities recognized the strategic necessity of defending Norway's reputation as a "green nation" while simultaneously evolving into a major petroleum power. During this period, Norway's production levels surged to rival those of OPEC nations such as Iran, Iraq, Kuwait, the United Arab Emirates, and Qatar.

Crucially, Norway's ascent as an oil state accelerated precisely when then-Prime Minister Gro Harlem Brundtland chaired the World Commission on Environment and Development, which introduced the goal of sustainable development in 1987. This occurred in tandem with the establishment of the Intergovernmental Panel on Climate Change (IPCC) in 1988, whose inaugural report in 1990 explicitly outlined the necessity of eventually eliminating all anthropogenic greenhouse gas emissions.

Norwegian Petroleum Policy: Immunized Against Climate Considerations

In December 2010, then-Minister of Petroleum and Energy Ola Borten Moe (Centre Party) made a landmark statement to the business newspaper *Dagens Næringsliv*. His comments sharply rebutted his cabinet colleague, Minister of the Environment Erik Solheim (Socialist Left Party). Solheim had suggested to his party's governing body that slowing the pace of opening new oil fields would be a primary lever for achieving Norway's target of a 20% emissions reduction by 2020. This debate occurred as the government was preparing a new white paper on climate policy.

Ola Borten Moe's response was unequivocal:

> The opening of new fields or the rate of extraction from existing wells will not be governed by climate considerations.

The revolving door between petroleum oversight and the industry itself provides a postscript. In 2015, Ola Borten Moe co-founded the oil company Okea. Between 2020 and 2024—during a period of intense climate debate—Okea was awarded numerous exploration and production licenses, including several as a designated operator on the Norwegian Continental Shelf. This "climate-blind" approach to petroleum management has proved remarkably consistent. Subsequent ministers from both the right-wing populist Progress Party and the Labour Party have issued nearly identical reassurances to the industry.

---

[2] Röstlund (2023).

In 1990, the Norwegian government, under the leadership of Gro Harlem Brundtland, established the CICERO Centre for International Climate Research. The institution describes its founding mandate as follows[3]:

> As a major exporter of oil and gas, Norway has a particular responsibility in international climate and environmental cooperation. On this basis, the CICERO Centre for Climate Research was established by the government in 1990.

In the early 1990s, researchers affiliated with CICERO played a pivotal role in providing the scientific and economic justification for a measure specifically designed to address the "Norway Paradox": the principle of an internationally cost-effective climate policy. This principle paved the way for the political consensus that for a high-income nation like Norway—already boasting a significant share of renewable energy—it is more cost-effective to fund emission reductions abroad than to implement expensive domestic cuts. The underlying logic was simple: to achieve the greatest possible emission reduction per Norwegian krone spent.

This principle eventually gained international traction, leading to the establishment of global mechanisms for emissions trading and compensatory measures. In her seminal work, *The Nature of Politics—The Politics of Nature*, the Norwegian environmental historian Kristin Asdal (2011) characterizes this development as a fundamental redefinition of the climate issue. She argues that the problem was transformed from a purely environmental matter into an economic issue, ultimately being reframed as an *oil issue*—a move that effectively protected the core of Norway's petroleum-driven economy.[4]

The Absurdity of Offsetting: When Reduction Becomes a Lost Opportunity

Over the years, I have been invited to numerous conferences. More often than not, these events are located far from Sogndal. From the organizers' perspective, the most practical solution is for me to fly. Since my presentations frequently address the necessity of changing consumption patterns, I have long felt the hypocrisy of boarding a plane to deliver a message about reducing one's carbon footprint.

Long before the COVID-19 pandemic made Zoom a household name, I began responding to invitations by asking to present digitally. While this usually works well with a bit of technical coordination, one particular encounter illustrates the fundamental absurdity of prioritizing compensation over actual emission reductions.

I was contacted on the phone by a manager of a large EU-funded project to speak on sustainable tourism at a seminar in a remote Swedish village. Reaching the venue would have required a convoluted itinerary involving multiple flights. I asked to present via a video link. After some hesitation, the organizer agreed,

---

[3] Alfsen and Moe (2011).

[4] Asdal (2011).

but the project manager remarked during the phone call, in a visibly irritated tone:

> In this project, we have a policy of offsetting all our flights. Because you are choosing not to fly here, we are missing a valuable opportunity to increase our efforts in carbon offsetting.

Another environmental policy measure based on a technological rationale is carbon capture and storage (CCS). While CCS is globally recognized as a necessary tool for eliminating residual greenhouse gas emissions to achieve net-zero targets, the technology holds a unique political utility in the Norwegian context. Specifically, it serves as a highly convenient mechanism to seemingly dissolve the "Norway Paradox." By shifting the policy focus from preventing emissions from occurring to managing them after generation, CCS unties climate mitigation from the necessity of halting petroleum production. In its ultimate consequence, this framework implies that a robust climate policy can co-exist with continued fossil fuel extraction. For a major oil-producing nation, this alignment is ideal, providing a powerful explanation for why Norwegian politicians across the spectrum have been so remarkably eager to champion CCS.[5]

In the following, I will describe a completely different story, namely that of Norwegian climate policy viewed from below—from a local political perspective. At times, this perspective offers entirely different views on what constitutes the problem and, consequently, how to solve it. More often than at the national level, alternative paradigms were permitted to emerge, advocating for profound changes in Norwegian society and our way of life. This is where my own and Western Norway Research Institute's (WNRI) climate research was situated from the early 1990s onward.

Municipalities Enter the Arena of Climate Policy

The emergence of a distinct local climate policy in Norway can be clearly traced back to the rise of local energy planning in the 1970s, driven by three converging factors: the 1974 oil crisis; the contentious debate over nuclear power and large-scale hydroelectric expansion; and the burgeoning environmental movement of the 1970s. The response to these debates was a paradigm shift toward local energy planning, significant attention to energy conservation (demand-side management), and a strong interest in diversifying renewable energy sources such as solar, wind, wave, and geothermal heat.[6]

The local energy planning of the 1970s replaced an earlier supply-oriented model rooted in the introduction of electricity in Norway in the 1920s, which aimed to ensure the delivery of electricity to consumers and national supply security. The

[5] Meadowcroft and Langhelle (2009).

[6] Sataøen et al. (2007).

new form of local energy planning thus represented an important shift in energy policy, directing attention toward consumption rather than production, and toward multiple forms of local energy use beyond just electricity. This pivotal shift toward consumption and away from production would occur once again—this time under the heading of local climate planning. This is a key turning point that I will describe further in this chapter.

In the early 1990s, major Norwegian urban municipalities began to systematically implement what became known as "integrated land use and transport planning." This initiative was anchored in a national policy guideline associated with the Planning and Building Act, with the aim of reducing transport demand. It was motivated, among other things, by the desire to lower greenhouse gas emissions. An important knowledge base for this new type of planning was the project Nature- and Environment-Friendly Urban Development (NAMIT), which involved myself and several colleagues at WNRI (see the discussion in the chapter on municipal environmental policy).

It was not until the mid-1990s, however, that a specific municipal *climate* policy began to take shape in Norway. This work was inspired by international trends in the late 1980s, which led to the establishment in 1990 of two different international alliances of climate-engaged municipalities: the Climate Alliance and the International Council for Local Environmental Initiatives (ICLEI). Three years later, the latter launched its own climate campaign called Cities for Climate Protection (CCP). In addition to supporting and subsequently promoting best practices for how municipalities can systematically work to reduce greenhouse gas emissions, these organizations also act as advocacy groups in international climate negotiations, often advancing demands that go further than those put forward by their respective national delegations.[7]

Two Norwegian environmental organizations, the Future in Our Hands (FIVH) and the youth branch of the Norwegian Society for the Conservation of Nature—Nature and Youth (*Natur og Ungdom* [NU])—were the first to develop local climate action plans in Norway. While awaiting public initiatives, the organizations began working on their own local climate action plans in 1995. FIVH developed a climate plan for Oslo, while around 20 local chapters of NU prepared plans for their respective municipalities.

The first local climate plan developed by a municipality was prepared by the City of Bergen, Norway's second-largest city, in the late 1990s, with formal initiation in 1997. This was the same year the Kyoto Protocol—the first international climate agreement—was adopted. Under Kyoto, Norway committed to ensuring that its national greenhouse gas emissions in the period 2008–2012 would not exceed 1% above 1990 levels. In February 2000, the City of Bergen adopted its climate plan with goals far more ambitious than the national commitment: Bergen aimed to reduce its CO_2 emissions by 20% by 2005, and its total greenhouse gas emissions by 30% by 2010, using 1990 as the reference year. This signaled that municipalities could—and

[7] Groven and Aall (2002).

should—take an active role independently of national directives, adopting targets that transcend rather than merely replicate state objectives.

Western Norway Research Institute Enters the Climate Policy Arena

In 1998, I participated in the first project at WNRI that focused specifically on local climate policy. This was an assignment for the Ministry of the Environment, where we were asked to survey international and Norwegian experiences with the development of local climate plans to strengthen the knowledge base for the ministry's work with Norwegian municipalities.[8]

The political backdrop for the assignment was the first Norwegian parliamentary report on climate policy—Parliamentary Report 29 (1997–98), "Norway's Follow-up to the Kyoto Protocol"—in which the role of local and regional authorities received little attention.[9] The report only briefly refers to Local Agenda 21 (see the chapter on municipal environmental policy), without describing the potential role of Local Agenda 21 initiatives within the national climate policy framework. However, the perception of the role of local authorities changed significantly as the government encountered political opposition to its proposed climate measures. In the summer of 1998, the Norwegian Parliament (the Storting) voted down the government's proposal for a general CO_2 tax, while a majority in the Storting's Committee on Energy and the Environment emphasized that "the central authorities have a particular responsibility to encourage counties and municipalities to develop local climate action plans."[10] In the autumn of 1999, after our report had been submitted, the Ministry of the Environment introduced a state grant, initially as a pilot scheme and, one year later, as a permanent scheme providing state funding for the development of municipal and county climate action plans.

From a situation in the spring of 1998, when local climate policy was virtually absent from the government's climate agenda, the government, in the winter of 1999, highlighted local climate policy as one of six prioritized policy areas in its climate strategy. It is therefore reasonable to question whether this represented a genuine strengthening of climate policy, or rather a weakening by shifting away from the "hard" instrument of the CO_2 tax to include the much "softer" approach of encouraging and financially supporting municipalities to develop local climate action plans. For us at WNRI, however, this shift in national climate policy provided an opportunity to conduct research on, and contribute to, the emergence of a new form of climate policy at the sub-national administrative level. Our small report became the entry point to an extensive research effort in the early 2000s, which included an evaluation of the state-initiated experiments with municipal climate planning that

[8] Groven et al. (1999).

[9] Ministry of the Environment (1998).

[10] Energi- og miljøkomiteen (1998).

I led,[11] several Research Council-funded projects in which I had the pleasure of participating, and a doctoral dissertation on local climate policy that I co-supervised with William M. Lafferty at the University of Oslo.[12] This also opened the possibility for us to become actively involved in processes that led to the development of a distinct consumption-oriented approach to climate policy—an approach promoted by key actors in pioneering municipalities such as Oslo and Stavanger and the Norwegian Association of Local and Regional Authorities (KS), but opposed and at times ridiculed by national politicians. More on this below.

"This is How the Politicians Want It!"

In 2000, the former Sogn og Fjordane County Municipality initiated work on its first climate plan, entitled "Regional Plan for Climate and Energy." This marked a pioneering effort. It was the first time in Norway—and potentially globally—that a regional authority developed a climate plan formally anchored in national legislation (the Norwegian Planning and Building Act).

WNRI was commissioned to develop the greenhouse gas inventory and propose mitigation measures. Simultaneously, the regional energy company, Sogn og Fjordane Energi (SFE), assessed power supply and demand, while the consultancy Energiråd AS evaluated energy conservation and efficiency measures.

During the final phase, the county authority invited the three parties to a coordination meeting to review each other's draft reports. We approached the peer review of our colleagues' work with the same scientific rigor we apply to our own research. At the meeting, Karl Georg Høyer, then Research Director at WNRI, delivered a blunt critique of the SFE draft:

> This draft would never pass the quality control protocols at our institute. The scientific quality is far too weak, and the conclusions appear politically motivated rather than scientifically justified.

To this, the representative from SFE replied disarmingly: "But this is how the politicians want it!"

The regional plan was eventually approved by the County Council in 2003 and submitted to the Ministry of the Environment for final ratification. The Ministry spent nearly two years reviewing the document, only to grant partial approval. Notably, the chapter on hydropower development was rejected. In its summary, the Ministry stated:

> The Ministry of the Environment approves the objectives and strategies in Chapter 3 (Greenhouse Gas Emissions) and Chapter 4 (Stationary Energy Use and New Renewable Energy Sources). However, the Ministry cannot unconditionally approve the sections concerning hydropower.

---

[11] Lindseth (2006a).

[12] Lindseth (2006b).

The Ministry pointed out that several of the county's objectives for new hydropower expansion were in direct conflict with national environmental interests.

This local push for expansion reflected a long-standing ethos. During the heyday of Norwegian hydropower development in the 1970s, Sogn og Fjordane was known as one of the most expansion-eager counties. A frequently cited remark by the local Labour Party representative, Sverre Lønningsdal, encapsulates this era: "Water that does not flow through pipes is wasted." This perspective was evidently still deeply embedded within SFE and continued to exert a powerful influence over the regional political environment.

From Experimentation to Normalization of the Municipalities' Role in National Climate Policy

In 2000, 7 million NOK were allocated in the national budget to support the development of municipal plans targeting both climate and energy. Following applications, this included 33 municipalities and 7 county authorities. WNRI, together with our close partner over several years, the Program for Sustainable Development (ProSus; see the chapter 1), conducted comprehensive broad-based studies of these initiatives in two phases: in 2002 (see Footnote 11) and in 2004.[13] Furthermore, we carried out three in-depth case studies of the cities of Stavanger[14] and Kristiansand (see Footnote 12), as well as my home county of Sogn og Fjordane.[15] In the studies of Stavanger and Sogn og Fjordane, we also participated directly in the planning process by preparing some of the background documents. Furthermore, we conducted a few international literature reviews: two on international networks for local climate policy in 1999 (see Footnote 8) and 2004,[16] and one on local climate planning and policy.[17]

The numerous and comprehensive studies of early examples of local climate planning provided the basis for two main conclusions:

- Many—but by no means all—municipalities wish to play an active role in climate policy, but this role must be flexible and adapted to differences among municipalities in terms of political priorities, local conditions, and institutional capacity.
- The orientation of local climate policy does not necessarily align with the orientation of national climate policy.

[13] Lindseth and Aall (2004).

[14] Farsund et al. (as listed in source) (2001).

[15] Groven (2001).

[16] Lindseth (2003).

[17] Teigland and Aall (2002).

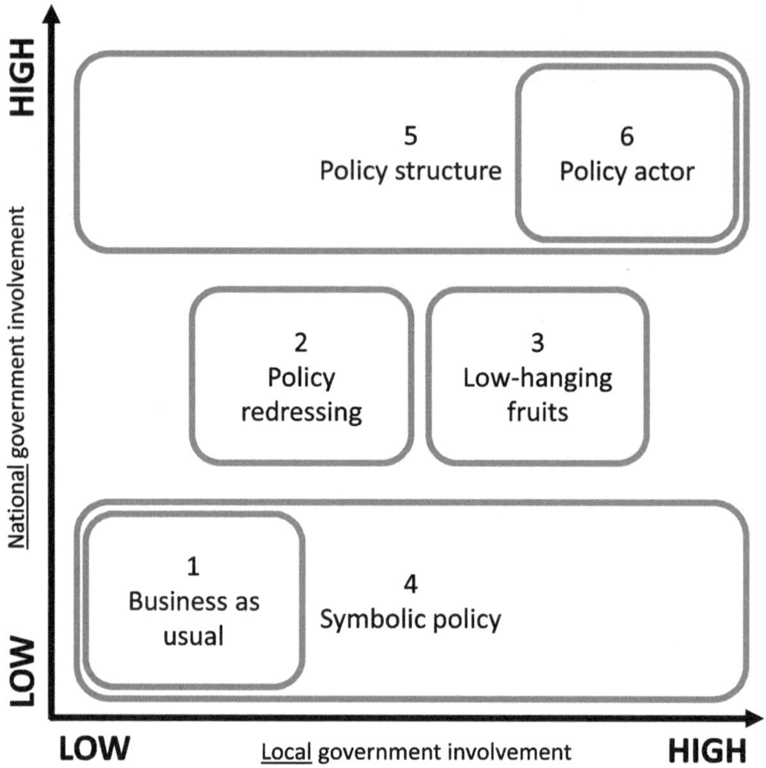

Fig. 5.2 The six-tiered ladder of local climate policy[18]

To understand how local authorities address the immense challenge posed by climate change, we must move beyond simple labels such as "success" or "failure." Our research provided the basis for developing a six-step ladder that describes the various roles municipalities assume in climate policy. This model illustrates the gradual—and often hesitant—acceptance of and commitment to climate policy at the local level and captures the mutual interdependence between national and local levels of governance (see Fig. 5.2).

The baseline phase is characterized by a lack of specific local initiatives and minimal national momentum. Our findings showed that Norwegian municipalities rarely prioritize climate change. When this study began, fewer than 10% of municipalities had committed to developing a specific climate policy. For the majority, climate action is simply not a political priority, and as a result, *business as usual* prevails.

The first example of local climate policy, and the *second* phase in our model, we called policy redressing. This step emerges at the intersection of moderate levels of local and national engagement and involves taking existing, well-established

[18] Aall et al. (2007).

programs—such as those designed to reduce local air pollution—and redefining them as climate initiatives. In Kristiansand, for example, a district heating system that had been under discussion for a long time was revived and ultimately implemented by linking it to the city's new climate action plan process. The measure itself was not new, but the climate context provided the necessary political weight and stakeholder support to bring it to fruition.

The *third* phase we called *picking low-hanging fruit*. It involves implementing measures that are profitable, uncontroversial, and easy to carry out, such as converting heating systems from oil or coal to cleaner electricity. Thanks to the extensive use of hydropower and affordable electricity in Norway, heating is already largely carbon-free. This forces many Norwegian municipalities to skip this comfortable phase, which municipalities in other countries have greatly benefited from, and instead go straight to the more challenging and "higher-hanging" fruit from the very beginning.

The *fourth* phase, which we referred to as, *symbolic policy,* is characterized by low national engagement, but may involve either low or high local engagement, and therefore comes in two versions:

- *Symbolism as distraction (low local effort)*: Here, ambitious goals, impressive planning documents, and calls for national cooperation primarily serve as a diversion to draw attention away from the lack of concrete action. Unfortunately, many global initiatives, such as the Local Agenda 21 processes, often fall into this "symbolic" trap.
- *Symbolism as a gateway (high local effort)*: Conversely, symbolic actions can serve as a gateway to genuine political innovation. Stavanger, for example, was a pioneering municipality in including indirect greenhouse gas emissions in its emissions accounting. This revealed that consumption-related emissions were nearly as large as direct local emissions, leading the city to implement measures to promote local and organic food in public cafeterias. Although the immediate reduction is small, the symbolic power of shifting the political agenda is significant.

The last two steps require strong support for local initiatives from national authorities. At this level, local governments serve as an effective structure for implementing national climate policy, promoted through a combination of national mandates ("stick") and financial incentives ("carrot"). The KS expressed a clear desire to the Storting for a more active state role towards municipalities in connection with the consideration of the second parliamentary report on climate policy[19]:

> The KS hopes that the understanding of the municipal sector's contribution to the Kyoto follow-up, as expressed in Parliamentary Report 54, is shared by the Standing Committee on Energy and the Environment. The KS therefore takes this opportunity to reiterate the municipal sector's desire to participate as a constructive partner in efforts to reduce Norway's greenhouse gas emissions.

The *highest* tier represents the most ambitious commitment, where local authorities act as *political actors* and go beyond the minimum standards and requirements

---

[19] Norwegian Association of Local and Regional Authorities (KS) (2002, April 19).

defined by the state. This is characterized by two main actions: setting more ambitious emission reduction targets than the national goals and introducing new issues that have not yet been addressed on the national climate policy agenda. Kristiansand exemplifies this by persistently raising the issue of regulatory measures to limit private car use—such as road pricing and parking reduction—a topic that has consistently been avoided in the national debate. Whether through global networks advocating for higher international ambitions (such as the Cities for Climate Protection campaign) or local leaders promoting controversial national measures, this level demonstrates that local authorities are not merely implementers but can serve as crucial drivers for the future climate policy agenda.

In a situation characterized by unclear national expectations, municipalities were largely left to develop the content of their local climate policies themselves. As a result, municipalities adopted an approach reminiscent of the local energy planning efforts of the 1970s. Measures for energy conservation, particularly those related to electricity consumption in buildings, therefore received significant attention. In a later phase of municipal climate work, this approach would come into direct conflict with national climate policy. In the official reporting of greenhouse gas emissions to the UN body that monitors national emissions, Norwegian electricity consumption is defined as having zero greenhouse gas emissions, since Norway is a net exporter of electricity generated exclusively from renewable sources—primarily hydropower, with some wind power in recent years. This means that measures to improve energy efficiency or reduce electricity consumption are not considered climate measures in Norway and are therefore excluded from national support schemes for local climate initiatives. In contrast, measures to increase the production of renewable energy receive substantial financial support from the state through various programs.

Many Norwegian municipalities have, since the pilot period in the early 2000s, developed energy and climate plans, which eventually came to be known simply as climate plans after the government decided to exclude electricity-saving measures from climate policy.

Over time, the national government's directives to municipalities have been strengthened, giving municipalities and county authorities a more clearly defined role in national climate policy. This includes the development of dedicated national guidelines anchored in the Planning and Building Act,[20] comprehensive guidance materials,[21] a "one-click" online service for presenting municipal and county-level greenhouse gas inventories,[22] and a national grant scheme for local climate initiatives which, since its inception in 2016, has totaled 1.7 billion NOK.[23]

[20] Ministry of Local Government and Regional Development (2024).

[21] Norwegian Environment Agency (n.d. a)

[22] Norwegian Environment Agency (2024).

[23] Norwegian Environment Agency (n.d. b).

A "Struck-Through" Policy: When the Ministry Almost Recognized the Municipal Sector

Between 2009 and 2010, the KS initiated several projects aimed at identifying how municipalities could systematically maximize their contribution to national emission reductions. One innovative proposal was the establishment of a municipal carbon credit system. Under this scheme, the state would pay the municipal sector for certified emission reductions, mirrored by the international mechanisms used to purchase credits from abroad. These reductions would then be reportable to the United Nations Framework Convention on Climate Change (UNFCCC).

The proposal gained significant traction and was included in the influential report *Climate Cure 2020: Measures and Instruments to Achieve Norwegian Climate Targets by 2020*. This report, published by a directorate group led by the Climate and Pollution Agency (2010), formed the central knowledge base for the upcoming white paper, Meld. St. 21 (2011–2012): *Norwegian Climate Policy*.

There was considerable anticipation within the KS as to whether their proposal would survive the transition from a technical report to official government policy. When the white paper was finally published, there was a moment of triumph: the proposal appeared to be included.

However, this excitement soon turned into profound disappointment. It emerged that the Ministry had accidentally released the penultimate version of the document. This draft contained a passage discussing the KS proposal, but it was shown as struck through in track changes mode—a literal visualization of the government's decision to delete the initiative. The error was swiftly rectified. A clean version was issued later that day, with all traces of the municipal quota system removed.

Saving Electricity and Salvaging the Climate—By Accident

As a researcher—or at least as a contract researcher—it is a rare and distinct pleasure to be genuinely surprised by the findings of one's own empirical work. I was fortunate enough to experience this during a project commissioned by the Norwegian Water Resources and Energy Directorate (NVE), published in 2012.[24] NVE was deeply puzzled as to why the aggregate energy consumption of Norwegian households had almost abruptly ceased to grow around 1990—a trend that flatly contradicted the agency's own forecasts at the time. While the Norwegian population and, crucially, its purchasing power had continued their steady ascent, direct residential energy use had decoupled from this growth. WNRI was tasked with solving this statistical mystery.

[24] Hille et al. (2012).

I had the privilege of working closely on this with two exceptional quantitative analysts: John Hille and Morten Simonsen, both of whom, to our great loss, passed away far too early.

What NVE expected, or perhaps hoped to discover, was that national energy and climate policies had finally borne fruit. On multiple occasions, the state had tightened building regulations under the Planning and Building Act to mandate energy-efficient solutions in new constructions (specifically in 1987, 1997, and 2007). Furthermore, following an electricity crisis during the winter of 2002/2003, the government introduced a direct subsidy of NOK 5000 for installing air-to-air heat pumps—an incentive utilized by nearly half of all Norwegian households. The dilemma for NVE, which tracked the exact real-time energy consumption of every single household in Norway, was that they possessed no data on end-use allocation. They simply did not know what the electricity was being used for. To crack this statistical nut, we adopted a dual approach. First, we aggregated all available ad-hoc surveys on end-use energy consumption conducted at various points in time. Second, we constructed a model that synthesized these discrete data points with aggregate electricity statistics. Based on assumptions about various drivers—such as changes in living space per household, appliance efficiency, annual demolition rates of existing housing stock, indoor temperature preferences, and actual outdoor temperatures—impacted consumption, the model generated a plausible distribution of energy use across different end-uses, complete with upper and lower estimates.

Our analysis was anchored in two stark realities: while aggregate temperature-corrected energy consumption in permanent Norwegian residences increased by a mere 3% from 1990 to 2009, it had surged by a staggering 55% in the preceding twenty-year period (1970–1990). The primary explanation for this structural shift, accounting for well over half of the divergence between expected and actual energy growth, was a marked slowdown in the growth of per capita residential floor space after 1990. Had residential space per person continued to expand between 1990 and 2009 at the same velocity as it did between 1970 and 1990, the total residential footprint in 2009 would have been 36% larger than it actually was. This stagnation was driven by a confluence of factors, among which demographic shifts—specifically the increase in what Statistics Norway defined as non-Western immigrant populations—proved to be the most significant. In 2001, this demographic group averaged approximately one-third less residential space per capita than the rest of the population. Another critical explanatory variable was the trajectory of housing prices and real interest rates, both of which had climbed since 1990. While Norwegians occupied roughly 66% more space per capita in 2009 compared to 1973, they paid an astonishing 600% more for their homes in real terms. Yet, what surprised both us and NVE the most was the remarkably marginal role played by deliberate energy and climate policies.

Our analysis of the 1990–2009 period revealed that political interventions designed to curb energy consumption—namely the revised building codes of 1987 and 1997, alongside other regulations, taxes, subsidies, and public information campaigns—likely accounted for only a minor fraction of the observed residential energy reductions. Instead, specific efficiency improvements were predominantly driven by autonomous, bottom-up initiatives undertaken by citizens themselves, independent of statutory mandates and devoid of financial subsidies. This was evident in the widespread retrofitting of insulation and windows in older housing stock, the replacement of water heaters, the installation of low-flow showerheads, and the overarching structural shift toward alternative energy carriers and heating technologies. Political decisions may indeed have exerted a powerful dampening effect on the expansion of per capita residential space after 1990; however, if they did so, it was primarily as an unintended consequence of macro-policies enacted for entirely different purposes, such as immigration control and monetary policy.

The Emergence of a Parallel Track to the State-Led Normalization of Local Climate Policy

What is measured matters. Rules for reporting national greenhouse gas emissions were established in the Kyoto Protocol of 1997 and follow the principle of territorial delimitation of national greenhouse gas inventories. This means that only emissions physically occurring within national borders are included. In today's globalized world, where consumer goods and production inputs increasingly cross-national boundaries, such territorially based greenhouse gas inventories end up heavily focused on emissions from the production process itself. Consequently, when compiling an inventory of greenhouse gases from production, climate policy naturally adopts the same focus—on production. As a result, climate measures related to Norwegian oil and gas production are limited to reducing emissions from the production process, for example by replacing gas-powered operations with electricity supplied via cables from the mainland to power platforms in the North Sea. In this way, Norway can eliminate the largest single source in its national climate accounts without having to address the scale of production, even though emissions from production are only a fraction of the emissions generated by the combustion of the same oil and gas in the countries to which Norway exports.

The territorial logic also applies in the national web-based system developed for the Norwegian Environment Agency for presenting municipal and county-level greenhouse gas inventories. As a result, emission levels vary greatly between municipalities. This is only to a very limited extent due to differences in population size, although the number of inhabitants varies considerably and therefore does influence emission levels to some degree (see Table 5.1). Oslo has a population 3625 times larger than the micro-municipality of Utsira, but only 100 times higher emissions. Calculated per capita, the relationship reverses (a ratio of 81 to 2). Area size tends to explain low emissions, as in Norway a large area generally corresponds to low population density, but by far the most important reason for variation in per capita emissions between municipalities is differences in *economic base*. The single-industry municipality of Årdal, where 97% of emissions originate from two aluminum plants, tops the list among Norwegian municipalities for emissions per person. Also ranking very

Table 5.1 Range of greenhouse gas emissions (total, per capita, and per km^2) compared to population size, land area, and municipality type (see Footnote 22)

Municipality	Emissions (highest = 100)			Population	Area (km^2)	Municipality type
	Total	Per capita	Per km^2			
Årdal	43	100	19	5200	977	Heavy industry municipality
Utsira	1	81	100	200	6	Small island and fishing community
Kautokeino	1	5	0.1	2800	9707	Agriculture and reindeer husbandry
Oslo	100	2	94	725,000	454	Capital of Norway, public administration

high is the small island municipality of Utsira, which is a major fishing community, where 92% of emissions stem from this economic activity.

If we shift our focus from a territorial to a consumption-based greenhouse gas inventory, the picture changes dramatically regarding local variation. The consultancy firm Ducky, with which we have collaborated for several years, has developed a service called "Folkets Fotavtrykk"[25] ("The People's Footprint"), offering a one-click web-based tool like that provided by the Norwegian Environment Agency for municipalities and county authorities (see Footnote 22). A free version of this tool, using data from 2022 distributed by county, shows that emissions per capita vary by plus/minus 10%—significantly less than the variation observed in territorially defined emissions inventories.

The differences between the county with the lowest per capita emissions (Oslo) and those with the highest (Innlandet and the three northernmost counties, which also have the most dispersed populations and the lowest population density) can be explained by two factors: mobility and the size of dwellings (and thus energy use for heating). For the other consumption categories used by Ducky—"goods and services," "food," and "emissions from public consumption"—the variations are practically negligible. What is important to note is that differences in per capita consumption-related emissions are substantially greater within than between municipalities, since this method of calculating emissions is inherently closely linked to economic factors—that is, the income level of individuals or households.

Interestingly, the oil city of Stavanger (see Footnote 14) and Oslo[26] were the first in Norway to develop such an alternative, consumption-based greenhouse gas inventory. I had the pleasure of participating in both projects. The first of these two projects was carried out in collaboration with my good friend and colleague John Hille at the Idébanken Foundation; the second with my equally good friend and colleague Professor William M. Lafferty at the University of Oslo and ProSus. Both

[25] Ducky (n.d.).

[26] Aall and Norland (2005).

projects demonstrated the dominant role of food, transport, and housing consumption in the overall climate footprint triggered by personal consumption (see also the chapter on sustainable consumption). This insight formed the basis for our later introduction of the alliteration "Biff, Bil, Bolig" (Beef, Car, and Housing) as a headline for the consumption areas that environmental and climate policy must crucially target. The two reports also showed that the impression one might get from territorial greenhouse gas inventories—that, per capita, it is more climate-friendly to live in cities than in rural areas—is incorrect. The political conclusion often drawn from this type of inventory, which shows that per capita emissions are lower in cities than in the countryside, is that centralization and migration from rural areas to cities are beneficial from a climate perspective.

However, a consumption-based climate inventory does *not* provide grounds for such conclusions. Emissions related to housing are indeed somewhat lower in large and medium-sized cities because a higher proportion of the population lives in row houses and apartment blocks, resulting in lower emissions associated with housing than in rural areas, where there is a much higher proportion of detached houses and thus greater living area per person. The same applies to everyday travel to work and daily shopping trips, where such journeys are more often made by public transport than by private car.

The opposite is the case for leisure travel. Residents of the largest cities have a much higher consumption of air travel, primarily due to greater proximity to airports that also offer international flights. Residents of the largest cities also have a higher frequency of long leisure trips by private car, typically trips "to the cabin," which are often located three to a four hours' drive each way from the city. In sum, this means that the difference in greenhouse gas emissions between urban and rural areas generated by private consumption is very small, albeit with somewhat *higher* emissions in the largest cities due to the significantly greater consumption of air travel compared to other parts of the country.

Over the course of the 2000s, the alternative consumption-based approach to local climate policy was incorporated into guidelines from the KS[27] and included as a dedicated chapter in the Norwegian Environment Agency's digital guide on local climate planning.[28] In both cases, a comprehensive model for developing greenhouse gas inventories is presented, distinguishing between *direct* and territorially limited emissions and *indirect* emissions resulting from household consumption of goods and services.

Norway's First Footprint Calculator: When the Climate Sceptic Had the Lowest Emissions

In 2000, in collaboration with John Hille at the Idébanken Foundation and commissioned by Stavanger Municipality, I took part in developing what would become Norway's first household climate calculator. This tool remained

[27] Oslo kommune, Hamar kommune, and Trondheim kommune (n.d.).

[28] Miljødirektoratet (n.d.).

available until recently (December 2025) available on the WNRI website. By answering 24 questions regarding household consumption across four categories—food, transport, housing, and leisure (including holiday homes and travel)—users receive data on their total greenhouse gas emissions, energy use, and ecological footprint compared to the national averages at the time.

To generate media interest for the launch, we invited members of the Stavanger City Council to calculate their individual footprints and consent to having their results published. The results provided a surprising twist: a representative from the Progress Party (*Fremskrittspartiet*)—a party that at the time vehemently opposed climate measures and remained deeply skeptical of climate change—recorded the lowest carbon footprint of all the council members. This irony, of course, sparked considerable attention in the local media.

HOPE for the Climate: Can Strategic Gaming Trigger Household Change?

In autumn 2013, I was invited to participate in a major European research proposal focused on strengthening consumption-oriented climate policy in high-consumption nations. The invitation came from Ghislain Dubois, the owner of a small French consultancy—Tec-conseil—specializing in social innovation within the climate sector. I had already collaborated with Ghislain on sustainable tourism, a partnership facilitated by my colleague Stefan Gössling. Now, Ghislain wanted me to join a large-scale experiment using strategic gaming as a research method. The idea was to build on the concept of the household climate calculator—like the one we had developed for the City of Stavanger a decade earlier—and supplement footprint calculations with a strategic game. This game would explore which lifestyle changes a household might be willing to undertake to align with the global goal of halving emissions by 2050.

In autumn 2014, the project received funding under the title Household Preferences for Reducing Greenhouse Gas Emissions in Four European High-Income Countries (HOPE). We immediately commenced the—as it would turn out—highly demanding task of recruiting participating families. The households were to be drawn from four cities: Bergen (Norway), Communauté de Pays d'Aix (France), Mannheim (Germany), and Umeå (Sweden). Although our initial ambition was to recruit 600 families per city, we eventually realized this was unfeasible. In fact, it fell to me to confront our quantitatively oriented German project leader, who had a background in social medicine, with this reality. We ultimately settled on 60–100 families per city.

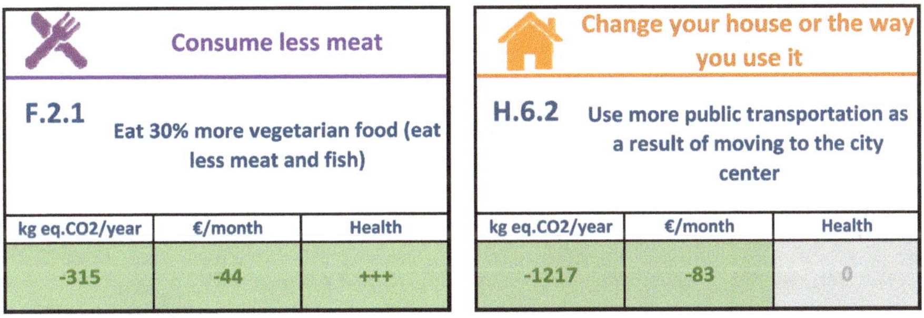

Fig. 5.3 Examples of consumption action card from the Housing and Food sectors in the HOPE household game

We invested significant effort in ensuring that the emission reductions associated with each consumption change were quantified as accurately as possible. The game followed a two-part structure[29]:

- *The Climate Calculator*: Households first completed an extensive online survey, answering roughly 150 questions regarding their consumption in physical or monetary units across four categories: transport, housing, food, and other goods.
- *The Board Game*: Armed with about 70 predefined "consumption cards" (see Fig. 5.3), a researcher acted as a game master during a 2-h session at the family's kitchen table. Each theme offered graduated options; for instance, dietary cards ranged from replacing 30% of meat and fish with vegetarian alternatives to going 100% vegetarian.

The two parts were interconnected; the deck of cards was tailored to each household's initial footprint. If a family already owned an electric car, the specific card proposing to buy an electric car was removed.

The board game was conducted in two distinct rounds. In the *voluntary round*, families ranked measures from 1 (easy to implement) to 5 (impossible). After negotiating which changes they could accept, we calculated the total reduction. Most families fell far short of the 50% reduction target, often to their own surprise, as they felt they had already made significant concessions. In the *mandatory round*, the premise shifted. Families were now required to select or exchange cards until the 50% target was met. Crucially, they were told to imagine that for every card chosen, a "magical" policy instrument was introduced, ensuring the measure applied to every household in the country.

Follow-up interviews revealed a psychological shift: when participants no longer had to act as "the odd green family on the street"—fearful of appearing "eccentric" to friends or neighbors—their willingness to change increased. The most notable shift was a significantly higher acceptance of reducing holiday air travel when they knew everyone else would be subject to the same restrictions.

[29] Hermann et al. (2018).

The HOPE project demonstrated widespread support for lifestyle changes to save the climate. However, it also proved that voluntary actions alone would, at best, bring society only halfway to the 2030 targets. To bridge the gap, there was general support for government intervention through robust policy measures to ensure collective action, making even the most contentious consumption changes socially and practically feasible.[30]

"Guilty, Me?": Putting Research to the Test on National TV

While the HOPE project was underway, I was contacted by producer Sigbjørn Holte regarding a four-part documentary series for NRK P3 (the youth channel of the Norwegian Broadcasting Corporation). Titled *Skyldig, jeg?* ("Guilty, Me?"), the series followed Holte himself as he spent four episodes attempting to discover whether it was possible to make his personal consumption truly sustainable.

In the first episode, Sigbjørn completed our online climate calculator before flying to Sogndal. There, my colleague Marta Baltruszewicz and I reviewed his results and provided guidance on how he could halve his emissions. The reality of the challenge set in immediately: for his return journey to Oslo, Sigbjørn was "forced" to take the night bus, while his cameraman enjoyed the luxury of a flight.

Subsequent episodes documented his lived experience of these lifestyle changes. He encountered significant social friction; for instance, when he hosted a vegetarian dinner for his friends, their lack of enthusiasm highlighted the cultural barriers to dietary change. In the final episode, Sigbjørn returned to Sogndal—this time by bus—to receive his final assessment. Despite his rigorous efforts, he had only managed a 25% reduction.

His conclusion was strikingly aligned with our research findings: to achieve a 50% reduction, an individual requires systemic support through regulations and taxes that compel everyone to change. Remarkably, Sigbjørn reached this conclusion through personal experience before we had even finished analyzing our project data.

[30] Dubois et al. (2019).

The documentary provides a compelling look at the practicalities of the "1.5°-lifestyle" and is available for viewing on the NRK platform.[31]

Producer Sigbjørn Holte discussing his personal carbon footprint calculation with researchers Marta Baltruszewicz and Carlo Aall at the Western Norway Research Institute in Sogndal. Screenshot from the NRK P3 documentary series "Skyldig, jeg?," reproduced with permission from NRK.

Is it Also Possible to Game your Way to an Alternative Local Climate Policy?

Inspired by the successful application of strategy games as a research method, I wanted to develop a follow-up game for municipalities where policy cards replaced consumption cards. In a Norwegian research project titled Local Transformation Towards a Low Emission Society (TRANSFORM), we began work in 2017 to develop such a game—this time designed as a Monopoly-inspired board game. Our team included professional game designers from Breda University of Applied Sciences, one of Europe's leading research environments for game mechanics.

Mirroring the HOPE project, we placed great emphasis on achieving the most precise possible quantification of emission reductions—in this case, generated by measures within the municipal jurisdiction. The game consists of 60 policy cards, each representing a specific intervention and containing five distinct data points:

---

[31] The documentary series *Skyldig, jeg?* [Guilty, Me?] was produced by Sigbjørn Holte and broadcast by the Norwegian Broadcasting Corporation (NRK P3) in 2021. The series can be accessed on the NRK platform via https://p3.no/skyldig-jeg/ (note: viewing is restricted to Norwegian IP addresses; accessed May 20, 2026).

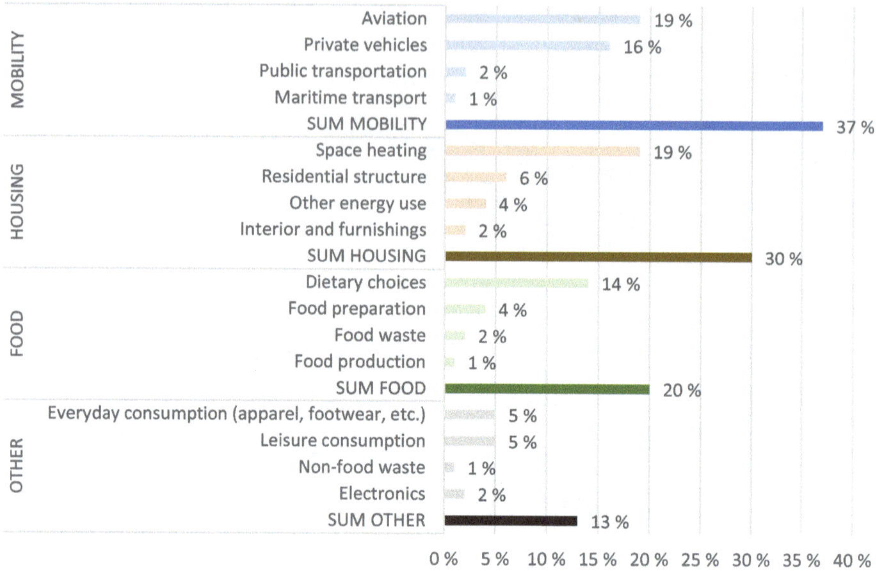

Fig. 5.4 Default relative distribution of consumption-related greenhouse gas emissions among four main and 16 subcategories of consumption in the TRANSFORM game

the concrete measure (e.g., introduction of parking restrictions), the magnitude and type of emission reductions, the implementation cost (requiring support from other players), a value representing the associated level of political conflict, and a long-term variable determining political support.[32]

We had a national reference group consisting of experts in municipal climate policy, and we invested considerable effort in helping them distinguish between the instruments they were accustomed to—addressing direct local emissions—and the consumption-oriented instruments that this game was designed to explore. Interestingly, when the game was conducted in the municipalities, municipal staff found the structure easy to navigate, perhaps because many participants worked in fields other than climate policy and were less bound by traditional silos.

The TRANSFORM game is based on an average Norwegian per capita emissions profile, with data sourced from the HOPE project and distributed across the same four main consumption categories (transport, housing, food, and other) and sixteen subcategories (see Fig. 5.4).

The game begins by presenting a *baseline emissions profile* and inviting participants to suggest adjustments based on local conditions. We then place tokens representing greenhouse gas emissions across the 16 categories on the game board (see Fig. 5.5).

In contrast to the HOPE game, where participants played as themselves, TRANSFORM is a role-playing game. Players are assigned one of five different roles

[32] Koens et al. (2020).

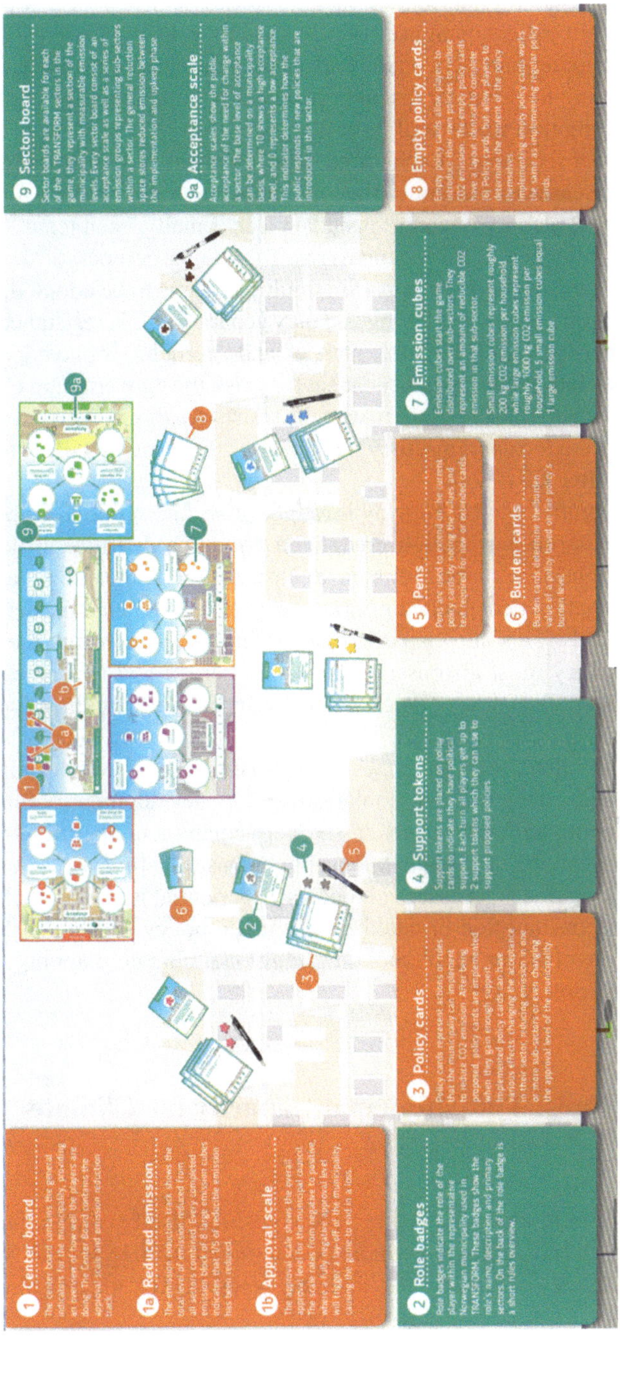

Fig. 5.5 The TRANSFORM game board. 1 = Current global warming prospect, 1a = Emissions reduced, 1b = Public support for the municipal council, 2 = Departments (teams), 3 = Policy card, 4 = Support token, 5 = Pen, 6 = "Burden" cards illustrating public support, 7 = Remaining emissions reduction potential per category, 8 = Blank policy cards, 9 = Emission categories, and 9a = Public acceptance level per sector

representing various municipal departments: Climate, Business Development, Planning, Technical Services, and Finance. Each team typically consists of 2–4 people. Crucially, teams are deliberately composed of players who, in their daily professional lives, work in other departments. The game was designed to be collaborative rather than competitive, aiming to stimulate discussions about the municipality's potential to influence the consumption patterns of its residents.

The game consists of six rounds. Each round begins with teams deliberating internally on which policy cards to present to the "municipal council." After each team proposes 1–3 measures, plenary negotiations take place. Finally, each team "votes" by using tokens to support their preferred measures. To simulate democratic decision-making, most policy cards require support from multiple teams to be adopted. Since societal transitions require political backing and may generate public resistance, each card is assigned numerical values for "acceptance" and "burden." If players choose radical measures associated with high resistance, they risk the "government collapsing"—meaning the council must resign because the resistance threshold is exceeded. Conversely, if players are too cautious, they risk "losing" the game by failing to meet the emission targets after six rounds.

This game also features a twist midway through, providing an obvious parallel to reality: there are simply not enough predefined policy cards to reach the 1.5° target. Midway through, once participants have mastered the game's logic and the balance between mitigation benefits and political resistance, the game master introduces blank policy cards. This allows participants to innovate and invent their own measures, for which they must also assign values for benefit and resistance. This element illustrates that radical new policies are necessary to achieve cuts in line with the 1.5° target.

In the scientific literature, serious games of this nature have been described as innovative tools for addressing wicked problems in policy development.[33] Based on experiences from playing with employees in six Norwegian municipalities and one county, as well as four cohorts of Climate Management master's students, my conclusion is that this type of game is an essential tool for experiential learning. It clarifies the components of a consumption-oriented local climate policy and highlights the strategic challenges involved in supplementing traditional climate planning with a consumption-based approach.

When the State Fails: The Municipal Capacity for Climate Action

Where, then, can future municipal climate policy position itself within a national framework that is so clearly failing? A report I co-authored in 2018 for the Norwegian Environment Agency demonstrates that the potential for municipalities to

[33] Haug et al. (2011).

reduce greenhouse gas emissions is significantly greater through a consumption-oriented approach than through the conventional territorial approach.[34] While the potential within territorially defined policy was estimated at 1.9–2.5 million tons of CO_2-equivalents (roughly 4–5% of Norway's 2018 emissions), the potential for an alternative, consumption-oriented strategy was estimated to be seven to ten times higher.

There are two primary reasons for the modest potential within conventional, territorially defined policy. *First*, there is little left to gain from reducing methane emissions from waste management—an area where municipalities have historically achieved massive cuts due to the 2009 national ban on landfilling organic waste. *Second*, emissions from heating buildings with fossil fuels have already been pushed toward zero following the 2020 national ban.

Consequently, municipalities are largely left with measures targeting construction activities, road traffic, and agriculture. However, even here, local agency is constrained: national mandates for transport electrification are already in motion, and agriculture remains a state-regulated industry with limited scope for additional local intervention.

In the autumn of 2023, the second official report on Norway's path to near-zero emissions by 2050 was published,[35] succeeding the first such report from 2006.[36] A fundamental shift between these two reports lies in their perspective on private consumption. While the 2006 report explicitly avoided proposing measures aimed at citizens' lifestyles, the 2023 report introduces the acronym "UFF" (in Norwegian "Unngå, Flytte, Forbedre" meaning "Avoid, Shift, Improve")—mirroring the international ASI framework as its primary hierarchy for mitigation. In a dedicated chapter on imported emissions related to Norwegian consumption—a topic entirely absent in 2006—the 2023 report states (p. 218):

> Norwegian consumption is extremely high by global standards and has a significant environmental footprint.

Based on this, it recommends that Norway establish a national target to reduce consumption-based emissions in line with the Paris Agreement. This represents a radical departure from the policy logic of the previous two decades.

The challenges of pivoting local policy toward consumption are, of course, significant. The first challenge—and arguably the least daunting—is technical uncertainty in reporting. How do we accurately quantify the impact of reduced meat consumption or fewer clothing purchases when the physical reduction occurs in another country?

The second, more formidable challenge is perceived public resistance to authorities intervening in private lifestyles. In my many lectures following the HOPE and TRANSFORM projects, I have concluded with a specific exhortation: politicians must now dare to "take the leap". This is a quintessential Norwegian metaphor derived from our tradition of ski jumping, a sport defined by a singular moment

[34] Westskog et al. (2018).

[35] Ministry of Climate and Environment (2023).

[36] Ministry of the Environment (2006).

of daring commitment. While Norway today has more abandoned ski jumps than active jumpers, the metaphor still resonates—it signifies the need for politicians to challenge their voters by "forcing" through the consumption changes necessary to meet climate targets. This is a conclusion that I can substantiate both through data from the HOPE project (see Footnote 30) and the practical experiences of municipal leaders during the TRANSFORM simulations. The leap may be daunting, but the evidence suggests that the public is more ready for collective action than politicians anticipate.

References

Aall, C., & Norland, I. T. (2005). *Det økologiske fotavtrykk for Oslo kommune* [The ecological footprint for Oslo municipality] (VF-report 2/02). Western Norway Research Institute/ProSus.

Aall, C., Groven, K., & Lindseth, G.. (2007). The scope of action for local climate policy: The case of Norway. *Global Environmental Politics*, *7*(2), 83–102. https://doi.org/10.1162/glep.2007.7.2.83

Alfsen, K. H., & Moe, T. (2011). *An assessment of Norway's national indicators for sustainable development* [Report]. Ministry of Finance/CICERO Centre for International Climate Research. https://www.regjeringen.no/globalassets/upload/fin/vedlegg/barekraft/cicero.pdf

Asdal, K. (2011). *Politikkens natur—naturens politikk* [The nature of politics—the politics of nature]. Universitetsforlaget.

Dubois, G., Sovacool, B., Aall, C., Nilsson, M., Barbier, C., Herrman, A., Bruyère, S., Andersson, C., Skold, B., Nadaud, F., Dorner, F., Moberg, K.R., Ceron, J.P., & Sauerborn, R. (2019): It starts at home? Climate policies targeting household consumption and behavioral decisions are key to low-carbon futures. *Energy Research & Social Science*, *52*, 144–158. https://doi.org/10.1016/j.erss.2019.02.001

Ducky. (n.d.). *Folkets fotavtrykk* [The People's Footprint]. Retrieved March 8, 2026, from https://folkets-fotavtrykk.ducky.eco/

Energi- og miljøkomiteen. (1998). *Innstilling fra energi- og miljøkomiteen om Norges oppfølging av Kyoto-protokollen* [Recommendation on Norway's follow-up of the Kyoto Protocol]. Stortinget.

Farsund, A., Hille, J., Aall, C., et al. (2001). *Klima- og energiplan for Stavanger kommune* [Climate and energy plan for Stavanger municipality] (Report RF 2001/184). Rogalandsforskning [Rogaland Research].

Groven, K. (2001). *Klimagassutslepp i Sogn og Fjordane: Historiske utsleppstal, framskrivingar og framlegg til tiltak* [Greenhouse gas emissions in Sogn og Fjordane: Historical emission data, projections, and proposed measures] (VF-report 6/2001). Western Norway Research Institute.

Groven, K., & Aall, C. (2002). *Lokal klima- og energiplanlegging. Norske kommunar som aktørar i klimapolitikken?* [Local climate and energy planning] (VF-report 12/02). Western Norway Research Institute.

Groven, K., Lundli, H.-E., & Aall, C. (1999). *Lokal klimapolitikk—internasjonale og nasjonale erfaringer* [Local climate policy—international and national experiences] (VF-report 4/99). Western Norway Research Institute.

Haug, C., Huitema, D., & Wenzler, I. (2011). Learning through games? Evaluating the learning effect of a policy exercise on European climate policy. *Technological Forecasting and Social Change*, *78*(6), 968–981. https://doi.org/10.1016/j.techfore.2010.11.011

Hermann, A., Amelung, D., Fischer, H., Wilkinson, P., Nilsson, M., Barbier, C., Dubois, G., Aall, C., & Sauerborn, R. (2018). Promoting health under the Paris Climate Agreement–Results from a European mixed-methods study, European Journal of Public Health, *28*(supplement 4), 170.

Hille, J., Simonsen, M., & Aall, C. (2012). *Trender og drivere for energibruk i norske husholdninger* [Trends and drivers for energy use in Norwegian households]. Report to the Norwegian Water Resources and Energy Directorate (NVE). VF Report 13/2012. Western Norway Research Institute.

Koens, K., Klijs, J., Weber-Sabil, J., Melissen, F., Lalicic, L., Hutchinson, K., Mayer, I., Önder, I., & Aall, C. (2020). Serious gaming to stimulate participatory urban tourism planning. *Journal of Sustainable Tourism*. https://doi.org/10.1080/09669582.2020.1819301

Lindseth, G. (2003). *The framing of an international concerted action for local climate policy: The case of the cities for climate protection* (VF-report 11/2003). Western Norway Research Institute.

Lindseth, G. (2006a). *Political discourse and climate change: The challenge of reconciling scale of impact with level of governance* [Doctoral thesis, Norwegian University of Science and Technology (NTNU)].

Lindseth, G. (2006b). *Politisk diskurs og klimaendringer* [Political discourse and climate change]. Presentation in connection with the Dr. Polit. dissertation, Norwegian University of Science and Technology (NTNU). VF Report 12/06. Western Norway Research Institute..

Lindseth, G., & Aall, C. (2004). *Kommuner og klima: En vanskelig kombinasjon. En spørreundersøkelse om klima- og energiplanlegging i norske kommuner og fylkeskommuner* [Municipalities and climate: A difficult combination. A survey on climate and energy planning in Norwegian municipalities and county municipalities] (Report 4/04). ProSus, University of Oslo.

Meadowcroft, J., & Langhelle, O. (Eds.). (2009). *Caching the carbon. The politics and policy of carbon capture and storage*. Edward Elgar.

Miljødirektoratet. (n.d.). *Beregne indirekte utslipp og livsløpsutslipp* [Calculating indirect and life-cycle emissions]. https://www.miljodirektoratet.no/ansvarsomrader/klima/for-myndigheter/kutte-utslipp-av-klimagasser/klima-og-energiplanlegging/beregne-indirekte-utslipp-og-livslopsutslipp/

Ministry of Climate and Environment. (2023). *NOU 2023:25: Omstilling til lavutslipp. Veivalg for klimapolitikken mot 2050* [NOU 2023:25: Transition to low emissions. Strategic choices for climate policy towards 2050]. Norwegian Government.

Ministry of Local Government and Regional Development. (2024). *Regulations on national climate and energy planning guidelines in the municipalities* (FOR-2024-12-20-3359). Lovdata. https://lovdata.no/dokument/SF/forskrift/2024-12-20-3359

Ministry of the Environment. (1998). *Report No. 29 to the Storting (1997–98): Norway's follow-up of the Kyoto Protocol*. Government Administration Services.

Ministry of the Environment. (2006). *NOU 2006:18: Et klimavennlig Norge* [A climate-friendly Norway]. Norwegian Government.

Norwegian Association of Local and Regional Authorities (KS). (2002, April 19). *Comments on Report No. 15 to the Storting (2001–2002): Concerning Norwegian climate policy* [Consultation letter]. Submitted to the Storting's Standing Committee on Energy and the Environment.

Norwegian Environment Agency. (2024). *Municipal greenhouse gas emissions: Statistics for Norwegian municipalities and counties* [Data set]. Retrieved March 8, 2026, from https://www.miljodirektoratet.no/tjenester/klimagassutslipp-kommuner/

Norwegian Environment Agency. (n.d. a). *Guidelines for the national planning guidelines on climate change adaptation* [Manual]. Retrieved March 8, 2026, from https://www.miljodirektoratet.no/ansvarsomrader/klima/for-myndigheter/klimatilpasning/veiledning-til-statlige-planretningslinjer-for-klimatilpasning/

Norwegian Environment Agency. (n.d. b). *Klimasats: Government subsidies for local climate action* [Policy brief]. Retrieved March 8, 2026, from https://www.miljodirektoratet.no/ansvarsomrader/klima/for-myndigheter/kutte-utslipp-av-klimagasser/klimasats /

Norwegian Offshore Directorate. (2024). *Key figures for the Norwegian shelf* [Data set]. Retrieved March 8, 2026, from https://www.sodir.no/fakta/data-og-analyser/nokkeltall/

Oslo kommune, Hamar kommune, & Trondheim kommune. (n.d.). *Veileder for klimabudsjett som styringsverktøy* [Guide for climate budgets]. Norwegian Association of Local and Regional Authorities.

Röstlund, L. (2023). *Norge-paradokset* [The Norway Paradox]. Cappelen Damm.

Sataøen, H., Aall, C., Mjøs, T., & Sijstermans, R. (2007). *Lokale energiutredninger: Evaluer-ingsrapport* [Local energy assessments: Evaluation report] (VF-report 5/2007). Western Norway Research Institute.

Statistics Norway. (2024). Table 13931: *Emissions to air, by source (detailed list), pollutant and year* [Data set]. Retrieved March 8, 2026, from https://www.ssb.no/statbank/table/13931

Teigland, J., & Aall, C. (2002). *Lokal klima- og energiplanlegging. Noen nyere nordeuropeiske og nordamerikanske erfaringer* [Local climate and energy planning] (VF-notat 2/02). Western Norway Research Institute.

Westskog, H., Selvig, E., Aall, C., Amundsen, H., & Jensen, E. S. (2018). *Potensial og barrierer for kommunale klimatiltak* [Potential and barriers for municipal climate measures] (M-981). Norwegian Environment Agency.

Chapter 6
Climate Change Adaptation as Failure or a Catalyst for Critical Climate Action?

Abstract This chapter examines the paradox of Norwegian climate change adaptation: a nation with unparalleled financial resources, yet a significant and growing maintenance debt in its physical infrastructure. By juxtaposing the costs of climate-related damages and necessary responses—such as landslide and flood protection—against the immense cash flows from the petroleum sector and the growth of the Government Pension Fund Global (GPFG), the author explores whether Norway is choosing reactive compensation over proactive resilience. The analysis reveals that the national maintenance backlog and required climate change adaptation measures (estimated at NOK 3560 billion) could be addressed by a marginal adjustment of the fiscal rule governing annual withdrawals from the fund of 0.25 percentage points. The chapter concludes by questioning whether the current adaptation strategy represents a failure to prioritize long-term safety, or whether the nation's financial "war chest" provides a unique, yet largely unexploited, opportunity for a transformative "shot in the arm" for critical climate action.

Introduction

When I was awarded professorial competence at Aarhus University in 2012, the faculty asked whether I wished to be titled Professor of Climate Policy or Professor of Sustainable Development. I chose the latter. Four years later, when the Western Norway University of Applied Sciences offered me a professorship, they asked the same question. Again, I chose sustainability over climate. Nevertheless, over the past decade my research has focused almost exclusively on climate—and specifically on climate change adaptation. In my professional life, I now call myself a climate researcher.

My colleagues' research and my own on climate issues emerged as a direct continuation of our work on local environmental protection and Local Agenda 21. In 1999, the Western Norway Research Institute (WNRI) received funding from the Research Council of Norway for a strategic institute program on Local Agenda 21, which helped finance my doctoral work. While mapping international experiences with

C. Aall, *Sustainable Development*, https://doi.org/10.1007/978-3-032-35930-8_6

Local Agenda 21, my colleagues and I documented the first examples from the early 1990s of municipalities recognizing the need to develop policies for adapting society to the consequences of climate change.[1] A core conclusion from that program was that climate change adaptation would—and should—be placed on the political agenda in Norway.

Surprisingly, few other research communities in Norway in the late 1990s anticipated this development. This foresight laid the foundation for the central role that WNRI and I would later play in research both *for* and *on* climate change adaptation in Norway. By the 2010s, we were almost unrivalled in this field, and I had the privilege of leading most of the research and development projects on this topic in Norway during that period. Over time, our work became increasingly academically oriented, leading to several major Research Council-funded and international projects, and ultimately to my current role as director of the Norwegian Centre for Sustainable Climate Change Adaptation (Noradapt). Below, I discuss this development in greater detail.

"It Was Surprising—But Actually Wise—That You Became so Deeply Involved in the Work on Climate Change Adaptation"

I have had my office at Campus Sogndal since its establishment in 1998. The campus represents a unique collaboration between the state (Western Norway University of Applied Sciences), the county (Sogndal Upper Secondary School), the municipality (Sogndal Lower Secondary School), the sports community (Sogndal Football), and several knowledge-based enterprises, including WNRI. This partnership has enabled a small town of 4500 inhabitants to host an 84,000-square-metre facility, serving 3500 students and 1000 employees.

A central driving force behind Campus Sogndal and former president of the Norwegian Football Association, Yngve Hallén—whom I have known since the Institute moved to the campus—recently revealed over a cup of coffee that he was initially surprised by my deep involvement in climate change adaptation. He had first seen the field as a way of protecting business-as-usual, which seemed to contrast with my established focus: turning business-as-usual into sustainable development.

However, he noted that after hearing me explain that climate change adaptation must take place within the framework of sustainable development, he realized it could be a strategic way to engage a broader audience in the necessary discussion on societal transformation. As he remarked to me:

> Everyone is concerned with protecting themselves against external threats, such as climate change, even those who are sceptical of environmental policy; but by first highlighting the consequences of climate change, it may become easier to involve these same people in discussions about the necessity of also changing societal development.

[1] Groven et al. (1999).

My own rationale for researching climate change adaptation is simple. If you dig deep enough into the measures required to adapt society to the unavoidable impacts of climate change—which will occur regardless of how quickly we reduce greenhouse gas (GHG) emissions—you inevitably arrive at the necessity of reducing economic growth in the wealthy parts of the world.

Climate Change Adaptation Enters the Climate Policy Agenda in Norway from the Sidelines

Climate change adaptation has undergone a maturation process from "absence" to "presence" on the climate policy agenda in Norway, as in most other countries, following the establishment of the UN Intergovernmental Panel on Climate Change and countries' endorsement of the UN Framework Convention on Climate Change. I argue that this process was characterized by deep institutional and political resistance, which can be categorized into five distinct phases: (1) ignorance; (2) acceptance—with significant support from Sogn og Fjordan; (3) conditional effort; (4) adaptation to natural hazard events; and (5) from natural hazards to climate risk, and addressing new forms of climate risk.

Phase 1: Ignorance

The first phase in Norway—which I define as spanning from the 1990s to the early 2000s—featured two major natural events that should have served as wake-up calls: the New Year's Hurricane of 1992 and the Vesleofsen flood of 1995. The former remains the most powerful hurricane on record in Norway. The latter was named after the Storofsen of 1789, the largest flood in Norway's recorded history. The name literally translates as "The Little Frenzy," contrasting with its eighteenth-century predecessor, "The Great Frenzy."

Despite these events, the early debate was characterized by a near-total lack of imagination. I am not referring here to "climate scepticism"—which implies active resistance—but rather to a passive lack of imagination. We simply believed that climate change was something that only affected "others," such as low-lying nations like Bangladesh or the Netherlands, but not us in Norway.

Who Has Heard of Tornadoes in Norway?

In the summer of 2000, the County Governor of Oslo and Akershus conducted an emergency exercise for municipal managers. The scenario involved an extreme weather event in Hurdal—specifically a strike by high-velocity winds; a tornado. During the evaluation, the participants dismissed the scenario as highly unrealistic.

However, the very next year, the same area was hit by an intense storm—likely a tornado—that flattened large forest areas and left thousands without power. The weather services had failed to predict it. Afterwards, the common refrain was that "no one could have imagined such an event, as it had not occurred in living memory." If the storm had struck just a little further south, in the capital city of Oslo, the consequences would have been devastating.[2]

Phase 2: Acceptance—The Sogn Og Fjordane Rebellion

In Phase 2, we see a remarkable institutional shift: national adaptation policy emerged from the civil protection sector rather than the environmental administration. While environmental authorities had dismissed climate impacts in their 1995, 1998, and 2001 white papers, the Ministry of Justice broke the silence in 2004. Their White Paper on Civil Protection (No. 39) explicitly warned that the strain on infrastructure and housing might exceed current levels due to increased flooding and landslides.[3]

Crucially, the Ministry of Justice pointed out that the solutions lay deep within the remit of the Ministry of the Environment—specifically in land-use planning and building regulations. This cross-sectoral challenge set the stage for a fruitful interplay between top-down initiatives and a bottom-up rebellion from my home county of Sogn og Fjordane.

As it turned out, the county of Sogn og Fjordane became the national engine for climate change adaptation, led by the proactive County Governor Oddvar Flæte and Emergency Preparedness Chief Haavard Stensvand. While national environmental authorities were still hesitant, local initiatives moved rapidly[4]:

- In 2005, Flora Municipality situated in Sogn og Fjordane became the first in Norway to adopt a formal adaptation measure.
- By 2008, the county had organized the first municipal gathering dedicated to climate change adaptation, and WNRI was commissioned to collect regional examples to inspire a national online portal.

[2] Husabø (2010).

[3] Norges offentlige utredninger (NOU) (2010).

[4] Husabø (2009).

- In 2009, the county of Sogn og Fjordane also pioneered the work on Norway's first regional climate plan with a dedicated climate change adaptation chapter.

This regional momentum forced the government's hand. In December 2008, an official expert committee was appointed to assess Norway's climate vulnerability and develop a national policy. It was only natural that Oddvar Flæte was appointed to chair this commission. The "Flæte Commission" presented its findings in 2010, providing the foundation for Norway's first *White Paper on Climate Change Adaptation* in 2013. Through this ten-year struggle—driven largely by regional persistence—adaptation finally integrated into national climate policy.

The Struggle Over Vulnerability Perceptions: When Norway Almost Became a "Climate Winner"

As part of the Flæte Commission's work, I participated in a research group consisting of experts from CICERO, economists from ECON Pöyry, and my colleagues at WNRI. The group was commissioned to synthesize existing knowledge on the potential impacts of a warmer climate on Norwegian society.[5] During this process, significant professional friction arose.

The majority—led by the economists—leaned toward socioeconomic modeling. They argued that Norway might emerge as a *climate winner*, suggesting that rising temperatures would strengthen agriculture and increase forest growth. At WNRI, we represented a critical minority. We raised two fundamental objections:

- *The Illusion of Isolation*: The majority's optimistic conclusions relied on the assumption that the indirect climate change consequences in Norway of climate change taking place in the rest of the world would be negligible. We found it scientifically irresponsible to assess Norway's vulnerability while disregarding the dramatic impacts occurring abroad.
- *Scientific Uncertainty*: We criticized the lack of emphasis on systemic and ecological risks that could not be captured by simplistic economic models.

To make our stance clear, WNRI published the final report on our own website with a clarifying comment, challenging the notion that climate change would be anything but dramatic for Norway. We pointed out that vulnerability analyses based on optimistic temperature projections—and a disregard for international "spillover" effects—were deeply flawed.

Our persistence paid off. The task of assessing the indirect climate risks—such as food security and refugee situations—was subsequently reassigned to the Fridtjof Nansen Institute, finally bringing these critical global perspectives into the Norwegian debate.[6]

---

[5] Aaheim et al. (2009).

[6] Buan et al. (2010).

Phase 3: Conditional Action—The Fear of Distraction

Even as adaptation gained acceptance, it faced an unexpected barrier: resistance from the environmental movement and ambitious climate politicians. Key actors feared that a strong emphasis on adaptation could divert attention from addressing the root cause: reducing emissions.

True Climate Activists Do Not Engage in Climate Change Adaptation!

In 2008—the same year the government established the Flæte Commission—I searched the website of Norway's largest environmental organization, the Norwegian Society for the Conservation of Nature (Naturvernforbundet). Using the terms "climate" and "climate change adaptation," I got 646 and 1 result, respectively. On the websites of other leading organizations, there were no results at all for "adaptation".

Even as late as December 2025, during a public hearing organized by the government-appointed Expert Committee on Climate Change Adaptation, the World Wide Fund for Nature (WWF Norway) opened their statement with a warning[7]:

> The climate crisis has already progressed to the point where we cannot avoid discussing and prioritizing adaptation measures. However, the most important means of limiting the impacts of climate change is to reduce emissions. Therefore, the committee must be clear... that adaptation cannot be prioritized over emission reductions.

The environmental movement clearly still exhibits a reluctance to fully address adaptation—thirty years after the IPCC recognized it as a necessary component of climate action.

A related rationale for limited action is the perception that climate change adaptation is low politics—a technocratic necessity rather than an arena for high-stakes political debate. In a conversation with me, conducted after she stepped down as leader of the Socialist Left Party (SV) and assumed the directorship at CICERO, Kristin Halvorsen argued that adaptation should remain low politics to avoid partisan conflict. I disagreed, and stated that if we insist on keeping adaptation at a low politics level, controversial issues of high politics—like shifting funds from new roads to maintenance—will be systematically excluded.

[7] WWF Verdens naturfond (2025).

Ambitious Climate Policymakers Do Not Engage in Climate Change Adaptation Either!

This lack of prestige became clear in 2013 when the Minister of the Environment, Bård Vegard Solhjell from the Socialist Left Party (SV), launched Norway's first *White Paper on Climate Change Adaptation.* Instead of a high-profile launch, the invitation to the press was scheduled for the same day as the government's presentation of the Revised National Budget—a move that predictably buried the climate report under a mountain of financial news.

At the time, I was a member of SV's environmental policy committee. At a subsequent party meeting, I had the opportunity to ask Solhjell why he had hidden this milestone in the history of Norwegian climate politics. He replied candidly:

> There is no political prestige in climate change adaptation; this is not the area where you make a name for yourself in national politics.

This prestige gap obviously remains a fundamental barrier to proactive resilience in Norway.

Phase 4: Adaptation to Natural Hazard Events

Talking about the weather is the safest way to start small talk with a Norwegian. The popular saying "There is no such thing as bad weather, only bad clothing" reflects a cultural pride in resilience, but it also masks a systemic failure to prepare our infrastructure for a changing climate.

Oddvar Flæte, in his capacity as chair of the Official Norwegian Report (NOU) committee on climate change adaptation, emphasized that half the work of preparing for a future climate is accomplished if society simply adapts to the current one. His prime example was the massive maintenance backlog. In 2010, this "maintenance debt" was estimated at over NOK 200 billion—equivalent to more than one-third of Norway's gross domestic product (GDP) in 2010.

While addressing current challenges makes logical sense, the Office of the Auditor General issued a serious criticism in 2022, revealing that only 38% of municipalities had assessed climate risks 50 years into the future.[8] We remained focused on the climate of today, rather than the climate of tomorrow.

[8] Office of the Auditor General of Norway (2022).

The Challenge of Prevention

In 2004, the County Governor of Sogn og Fjordane conducted an emergency preparedness exercise in Stryn. Two of my colleagues participated as observers. The theme was extreme weather events, but with the added premise that their frequency and intensity would increase due to global warming. Stryn was chosen because of its vulnerability to glacial melt from the Jostedalsbreen glacier, potential flooding of the Stryneelva river, and landslides on the surrounding steep slopes.

What struck us was a clear bias toward post-event response. Whenever a scenario required strengthening emergency services or crisis management, it received full and diligent attention. However, when pre-event prevention was introduced—for instance, questioning why a building had been constructed in a flood-prone zone in the first place—the participants remained focused on how to save the structure, rather than why it stood there.

We documented this post-event bias in a later study, where we interviewed 19 regional emergency managers about their needs in a changing climate. The results were telling: 77% of their statements concerned post-event crisis management, while only 22% addressed the need for enhanced pre-event prevention measures.[209]

Phase 5: From Natural Hazards to Climate Risk

In 2018, a government green paper on climate risk shifted the conversation from physical hazards to a broader understanding of risk, categorized as follows[9]:

- *Physical climate risk*: Direct risks triggered by changes in the physical climate.
- *Transition risk*: Risks arising from shifts in climate policy, technological developments, and market changes during the transition to a low-emission society.

The report critically highlights that climate-related events occurring in other countries can also affect Norway through international trade, financial instability and migration—later termed *transboundary climate risks*. Indeed, the report concludes that these indirect climate risks likely represent a greater challenge for Norway than the physical risks manifesting within its own borders.

This shift from "vulnerability" to "climate risk" was reflected five years later in the second *White Paper on Climate Change Adaptation*, where the term climate risk occurred 17 times, compared to only twice a decade earlier. This mirrors a profound transition in the international climate discourse. Data from the IPCC assessment

[9] Norges offentlige utredninger (NOU) (2018).

reports show that while vulnerability was the dominant term in 2007,[10] it has been superseded by climate risk since 2022.[11]

In 2024, WNRI was commissioned to develop a new methodology for national assessments.[12] Although the authorities retained the legacy term vulnerability in the title of the report, the framework is, in practice, a comprehensive risk analysis encompassing physical, transboundary, and transition risks.

Research "for" Climate Change Adaptation

While the Research Council of Norway and academic circles debated the role of social sciences in climate research, several municipalities initiated their own research and development projects. WNRI assumed a key role in this bottom-up movement. Before my research career, I served as a municipal environmental manager—an experience that provided me with a unique professional network to establish municipal "laboratories." Here, we could develop and test new methods through action research.

The Struggle for a Scientific Agenda

As adaptation gained political traction, a struggle emerged to define what adaptation research entailed. Initially, the field was dominated by the "hard" sciences—meteorology and physics where success was measured by the precision of climate models. To entice more social scientists into the field, the Research Council of Norway organized what it termed "recruitment conferences" for the social sciences in 2004 and 2005.

At the second conference, which I attended as a young researcher alongside my colleague and mentor, Professor William M. Lafferty, a fascinating exchange took place between two "titans" of Norwegian academia. Anton Eliassen, the Director of the Norwegian Meteorological Institute, attempted an inclusive tone by stating: "I have nothing against normative research." The response from Alf Inge Jansen, a towering figure in political science, was sharp: "The reason many of us in the social sciences have been reluctant to get involved is that we perceive the current orientation of climate research as already being highly normative."[13]

Jansen's point was profound: the very choice of what to model and how to define risk is a normative act. This academic hesitation from the established universities made it easier for independent institutes like WNRI to secure key research contracts in the burgeoning field of adaptation.

[10] Intergovernmental Panel on Climate Change (IPCC) (2007).

[11] Intergovernmental Panel on Climate Change (IPCC) (2022).

[12] Dannevig et al. (2024).

[13] Aall (2005a).

In 2000, research dedicated to adaptation in Norway was almost negligible. To bridge this gap, a pre-programme was launched in 2002, with WNRI leading four foundational studies that anticipated the future of the field:

- *The 1992 New Year's Eve Hurricane Study*: This 2002 project was the first empirical investigation in Norway of a major weather event viewed through a climate change lens—focusing on what remains the most destructive wind event in modern Norwegian history. To measure societal resilience, we used public statistics to track long-term socioeconomic effects ten years after the storm, comparing the 30 most affected municipalities with 30 most-similar control municipalities. We analyzed population trends, local economy, and employment. The only statistically significant difference was a spike in birth rates nine months after the event in the affected municipalities—a "baby boom" echo often observed after similar disasters worldwide. Our conclusion, later confirmed by numerous studies, was that Norwegian society possesses remarkably high resilience to natural hazards.[14]
- *Tourism and Climate study*: An international literature review addressing how weather and climate influence travel and recreation behavior. It identified a glaring lack of research on the interplay between changes in the natural environment and climate on travel patterns. The study laid the groundwork for several major projects that I later led, aiming to fill these gaps.[15]
- *The Four-Pillar Risk Framework development*: This 2003 study established the institutional core of our research, focusing on four key institutional systems: civil protection, natural hazards insurance, municipal planning, and environmental management.[16] Our analytical framework—integrating climate impact, environmental vulnerability, societal exposure, and local response capacity—remarkably anticipated the four-pillar risk framework adopted by the IPCC twenty years later in its *Sixth Assessment Report* (hazard, vulnerability, exposure, and response).[17]
- *Vulnerability Indicator development*: This final study in the pre-program launched a long-term quest for a functional indicator system to measure progress in adaptation.[18] We identified 16 indicators distributed across a tripartite structure: physical/biological, socioeconomic, and institutional vulnerability. This structure has proven scientifically robust, serving as the foundation for the Noradapt system used to rank all Norwegian municipalities by their level of physical climate risk first presented in 2023.[19]

[14] Teigland (2002b).

[15] Teigland (2002a).

[16] Aall and Groven (2003).

[17] Reisinger et al. (2020).

[18] Aall and Norland (2003).

[19] Rød et al. (2025).

What Is a Climate Scientist?

Many of us in the climate debate have received our share of misguided social media comments, but occasionally they become almost comical. In 2021, NRK published a story about a record drought in my home county, featuring a 74-year-old farmer standing by his dried-up river, stating he had never seen anything like it. I provided a comment for the article, citing the recently released *IPCC Sixth Assessment Report*:

> Climate change is here now. This is not a matter of natural variations... We must prepare for more extreme weather, including intense heatwaves and prolonged droughts.

A few days later, I was contacted by the editor of a website called Værstat, true to its name (short for Weather Statistics), merely aggregates raw weather data,[20] He politely asked how I defined "climate scientist." I explained that the term encompasses everything from atmospheric physics to the social sciences and engineering required for climate change adaptation.

Next, I received an email from the news editor at NRK, informing me of a formal complaint from Værstat regarding "poor source usage." The complaint asserted that I was not a climate scientist, lacked the academic background to comment, and "strongly promoted political opinions" such as a managed phase-out of the oil sector.

NRK's response was firm, concluding that I was a highly relevant source who based his assessments on established IPCC reports. Curious, I investigated the complainant. The editor who had challenged my academic credentials to speak on climate change introduced himself as a graduate in business administration from BI Norwegian Business School. Ah, the irony!

Flora: Implementing Norway's First Climate Change Adaptation Measure

Flora, a coastal municipality in Western Norway, became my first "laboratory." In 2005, the head of planning, Øyvind Bang-Olsen, invited WNRI to assess the risks of sea-level rise and increased wind stress. This collaboration was pioneering: it produced the first policy measure in Norway justified specifically by climate change—six years before the national government released its first *White Paper on Climate Change Adaptation.*

---

[20] Verstat (2026).

Our research in Flora addressed a fundamental systemic question: is the problem "bad weather" or "poor planning"? We identified several weaknesses in the Norwegian planning system, such as liberalized regulations that reduced municipal oversight and a lack of geographic differentiation in national building codes. Put simply, building the same way in a sheltered inland valley as on a storm-exposed coast was no longer sustainable.[21]

The 44 cm Requirement and the Dagmar Reality Check

In 2007, Flora municipality adopted a plan for the new Botnaneset industrial area. For the first time in Norway, a legal requirement was incorporated to account for a projected sea-level rise of 44 cm over the next century:[211]

> All new quays and paved surfaces within the planning area must have a top elevation no lower than 322 cm above chart datum.

This decision was controversial. Raising the terrain required significantly more fill material than originally planned. The Chief Municipal Executive at the time, Fredrik Guldbrandsen, later remarked that they "could have saved millions" by ignoring the projection. The companies purchasing plots were equally sceptical of this "climate bill."

The turning point came on Boxing Day 2011, when the storm Dagmar hit. It was one of the most destructive storms in fifty years, bringing hurricane-force winds and a record-breaking storm surge. Across the region, boathouses were crushed and quays were submerged.

In Florø, the sea level reached 232 cm above chart datum. While neighboring areas suffered devastating damage, the industrial area at Botnaneset stood safe and dry. The earlier complaints from business owners vanished overnight, replaced with gratitude. The "unnecessary" investment saved them from far greater losses, proving that proactive adaptation is not a cost but an insurance policy for the future.

Fredrikstad: The First Integrated Climate Plan in Norway

In 2006, my colleague and friend Rolf Petter Heidenstrøm—then the municipal environmental manager in Fredrikstad—invited WNRI to introduce the concepts of climate vulnerability and adaptation to the municipal administration. While Fredrikstad is a large city in a Norwegian context, its challenges were universal: a mix of river flooding from the Glomma and rising sea levels in the harbor.

[21] Groven (2005).

Based on our report, Fredrikstad adopted a pioneering climate plan in 2007. It was the first in Norway to include a dedicated chapter with specific goals and measures for adaptation, focusing on three critical areas[22]:

- *River flooding*: Incorporating 50- and 100-year future scenarios for development along the river Glomma.
- *Sea-level rise*: Introducing strict requirements for minimum floor elevations for new buildings in port and riverfront areas (50–80 cm projected rise).
- *Stormwater management*: Early adoption of what we now call nature-based solutions, emphasizing open floodways.

Thus, Fredrikstad became the first municipality in Norway to adopt a climate plan with a dedicated chapter and specific goals and measures addressing climate change adaptation.

The World's First Scenario Model for Local Climate Vulnerability

From 2007 to 2011, the Norwegian Association of Local and Regional Authorities (KS) funded two national research projects involving eleven municipalities. I had the privilege of leading both.

The first project, titled "Storms, Landslides, Floods, and Oil Spills", was sparked by a devastating flood in Lom in 2004. A rainstorm combined with extreme glacial melt had caused the Bøvra River to overflow, damaging the village center. This prompted the Mayor of Lom to request a national project to help municipalities prepare for a changing climate.

WNRI was awarded the contract, and we brought in Tor Selstad, a leading expert in scenario methodology. His approach was groundbreaking, built on a simple but profound recognition: to understand local climate risk, we must understand vulnerabilities—i.e. how society develops—not just the climate itself.

For each of the seven participating municipalities (Hammerfest, Stjørdal, Ørland, Kristiansund, Lom, Tinn, and Fredrikstad) we developed both climate and societal scenarios. The model distinguishes between five local drivers of climate vulnerability[23]:

- *Population*: Increasing exposure and demographic shifts (e.g., an aging population).
- *Employment and industry*: The state of primary industries; for instance, declining forestry can increase flood risks.
- *Settlement and infrastructure*: The location of new housing and the state of the maintenance backlog.

---

[22] Sataøen and Aall (2006).

[23] Selstad (2008).

- *Mobility*: The steady increase in movement heightens overall societal vulnerability.
- *Mentality*: Perhaps the most complex driver—how local culture and attitudes affect the will to adapt.

Tor Selstad addressed the sixth driver—mentalities—of the participating municipalities as follows (p. 70)[24]:

> By 2025, Fredrikstad had become a haven for early retirees, thanks to its proximity to Oslo, the sea, and outdoor recreational opportunities. Hammerfest, on the other hand, has benefited from national wealth, which has bought many advantages, including expansive development along the waterfront and a focus on culture and attractions. In contrast, the rural communities of Tinn and Lom stand out. Lom is largely characterized by a culture of preservation, continuing a tradition of pride in the local language, cuisine, and people Tinn, by contrast, is an industrial community that has endured. Here, even in 2025, the culture and expertise of industrial workers have been preserved.

The feedback was overwhelmingly positive. Municipalities appreciated that the analysis placed equal emphasis on climate and societal developments, providing a holistic view rather than just a set of weather statistics.

From Climate Vulnerability Analysis to Climate Change Adaptation Strategies

In 2010, the same year the Norwegian government presented its first official report on climate change adaptation, WNRI—in collaboration with the Bjerknes Centre and SINTEF—was awarded a major project that served as a follow-up to our initial KS-funded study. The project partnered with six municipalities (Fredrikstad, Bergen, Voss, Melhus, Lyngen, and Tromsø) to develop local adaptation strategies that could serve as models for all Norwegian municipalities. This initiative marked the beginning of a long series of analyses regarding institutional barriers to municipal adaptation.

Our approach to developing a model for local climate change adaptation was rooted in a core recommendation of the World Commission on Environment and Development, which advocated a shift from effect-oriented to cause-oriented environmental policy within a new sustainability framework. As shown in Fig. 6.1, we applied this principle by linking it to the distinction between natural and societal climate vulnerability. This analytical model serves as the theoretical heart of our method. It emphasizes that while an effect-oriented strategy (e.g., building a flood wall) is necessary, it is merely reactive. A truly sustainable adaptation strategy must be cause-oriented, addressing the underlying societal drivers of vulnerability (e.g., reforming land-use policies). Crucially, the model also explicitly includes the barriers that often prevent a clear vulnerability analysis from being translated into a strategic response.

---

[24] Groven et al. (2008).

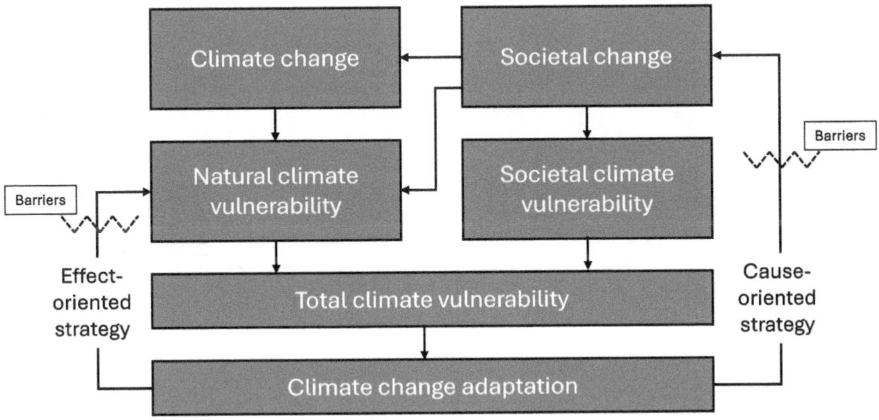

Fig. 6.1 Conceptual analytical model showing the relationship between climate change, societal change, total vulnerability, and adaptation strategies[25]

The issue of *uncertainty* has always been central to climate research and the broader climate debate. Some argue that the question of uncertainty is given too much emphasis, as Professor Lyla Mehta and colleagues observed (p. 8)[26]:

> Uncertainty needs to be understood … as a concept that can be strategically created and utilized by different actors. Whether one should attempt to reduce uncertainty or not should be regarded as part of highly political processes.

In my lectures over the years, I have always used this quote as a starting point. At the same time, there are inherent uncertainties in how climate change manifests at the local level. In our research projects, we have emphasized communicating this uncertainty directly rather than hiding it behind statistical averages.

In the above-mentioned project for the KS, my colleagues from the Bjerknes Centre employed projections from four different coupled atmosphere–ocean global models.[27] While combining models enhances robustness, the results often diverge. We spent considerable time evaluating how to communicate this to municipalities. We decided not to present average values; instead, we showed the range between the highest and lowest values from the models.

Our message was that the upper and lower estimates should be regarded as equally probable for practical planning. This approach met with objections from the Norwegian Water Resources and Energy Directorate (NVE). However, as shown in Fig. 6.2, the official presentation of climate scenarios in Norway has since moved steadily in our direction, de-emphasizing single mean values in favour of displaying the full uncertainty interval.

---

[25] Aall (2011b).

[26] Mehta et al. (2001).

[27] Miles and Richter (2011).

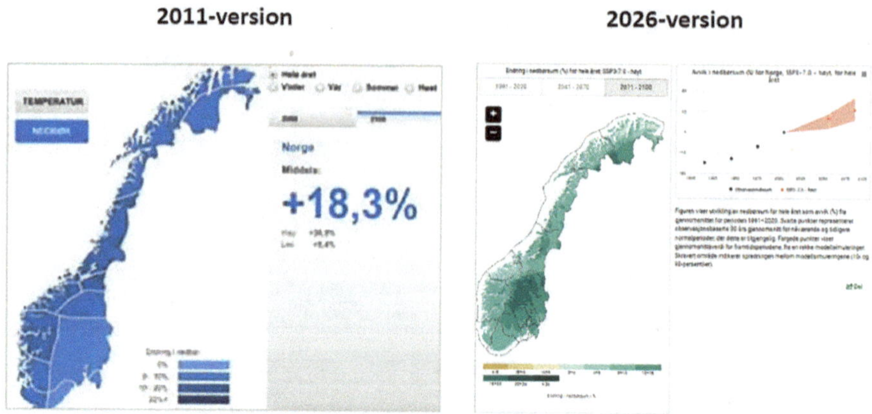

Fig. 6.2 Comparison of official climate scenario presentations. The 2011 version (left) focused on mean values,[28] while the current portal (right) emphasizes the uncertainty interval and model spread[29]

"We Feel Deceived"

In January 2011, I was invited to address officers and representatives from the Bergenshus Home Guard (HV-09) alongside the County Emergency Manager, Haavard Stensvand. As a former pacifist and conscientious objector, I found a certain irony in presenting to an audience in "green uniforms." However, this meeting would become a symbolic turning point in my understanding of risk communication.

During my talk, I warned against focusing solely on average values when projecting climate change, emphasizing instead the importance of communicating the full range between best- and worst-case scenarios. This point led to immediate nodding and murmuring among the officers. One of them spoke up with a bluntness that cut through the academic jargon[30]:

> We are trained precisely to plan based on worst-case scenarios. It is only natural that we do the same when addressing climate change. If we are only provided with the averages, we feel deceived.

This encounter made it clear that while climate science had spent years trying to "domesticate" uncertainty to make it palatable for policymakers, those tasked with managing the actual consequences were asking for the exact opposite.

[28] Norwegian Centre for Climate Services (NCCS) (2011).

[29] Norwegian Centre for Climate Services (NCCS) (2025).

[30] Aall (2011a).

It confirmed that our alternative approach to communicating uncertainty—presenting the full spread of possibilities—was, in fact, the most realistic path forward for a society in transition.

In our work with the six pilot municipalities, it became clear that the primary challenge for local adaptation did not necessarily concern future projects. To move local government from reactive "firefighting" to strategic resilience, we developed a set of prioritized criteria. These principles were designed to translate the project's goal of a cause-oriented adaptation policy into practical reality:

- *Adapt to the current climate first*: Address existing structural weaknesses before planning for future extremes.
- *Prioritize no-regret measures*: Focus on actions that are justifiable under all likely climate scenarios and do not conflict with broader sustainability goals (e.g., GHG emissions).
- *Prioritize strategic over operational work*: Climate change adaptation must be integrated into long-term land-use and financial planning, rather than being treated as a series of isolated technical projects.
- *Shift from expansion to preservation*: Address the "growth logic" by prioritizing the massive maintenance backlog of maintenance for existing infrastructure over new public construction.

The final point—the need to prioritize existing physical infrastructure over new growth—proved to be the most politically sensitive, yet most necessary, step toward achieving genuine robustness.

The Headline that Shook the Municipal Sector

When we conduct commissioned research, the client often wishes to issue a press release upon completion. Usually, we prepare a draft for their approval. This was the case for our project on municipal and county infrastructure.

Working with the project manager at the KS—my good friend and colleague Ole Jørgen Grann—I prepared a draft that highlighted our recommendation for a "construction halt." I knew this was a radical proposal for an organization like the KS—which represents the very owners of that infrastructure—to promote. I expected it might be toned down as it moved up the chain of command.

To my surprise, after several internal rounds at the KS, the press release was accepted in its entirety. The "construction halt" was not only included but was even used as the lead headline in the KS's own newspaper, *Kommunal Rapport*. It was a powerful confirmation that the scale of the maintenance crisis was finally being recognized at the highest political levels of local government.

As the reactions to the call for a construction moratorium demonstrated, we had struck a nerve within the local government sector. Our data confirmed an uncomfortable reality: much of the physical infrastructure required to withstand the climate of 2050 or 2100 has already been built. Simultaneously, we have expanded our physical footprint at a rate that far outpaces our capacity to maintain it.

In a changing climate, an unmaintained asset is a liability, regardless of how robustly it was originally constructed. Future resilience, therefore, is not primarily about adding more layers of protection or safety zones for new projects. Instead, it is about how society can secure its existing foundations. To find the resources for this monumental task, we argued that society must find a way to halt the inherent drive for constant expansion. This realization required a strategy that addressed the root cause of vulnerability: the growth logic itself.

The Reality Gap: Infrastructure Expansion Versus Maintenance

Despite the brief political spotlight on a construction moratorium, the subsequent public debate largely sidelined the radical idea in favour of a "yes, please, both" approach. Official policy became a dual desire: increasing funding for maintenance while simultaneously maintaining high levels of investment in new infrastructure.

Figure 6.3 illustrates the consequences of this policy using data compiled from the Norwegian Association of Consulting Engineers (RIF), KOSTRA (Statistics Norway), Norsk Vann, and the National Transport Plan. To measure the expansion of our built environment, we use replacement value (A) as a proxy for the total physical volume of infrastructure. When this value increases, it reflects a society that is continuously adding new assets—and thus new maintenance obligations—to its balance sheet.

These figures, while based on rough estimates, reveal two clear and concerning trends:

- *Expansion outpaces maintenance*: The total volume of physical infrastructure (measured by replacement value) has increased more rapidly since 2010 than the annual expenditure on its operation and maintenance.
- *Growing maintenance backlog*: Despite increased spending, the backlog has continued to grow across most sectors. In the national road sector (2011–2023), for example, over 130 billion NOK (constant 2011 values) were spent on new investments, while less than 50 billion NOK went to operation and maintenance.

The resulting 25 billion NOK backlog (in 2011 NOK) is a direct result of political priorities. In theory, this backlog could have been eliminated. Had the ambition for new road projects been reduced by just 15–20% annually—with those funds redirected toward renewal—Norway would have a national road network today with zero maintenance backlog.

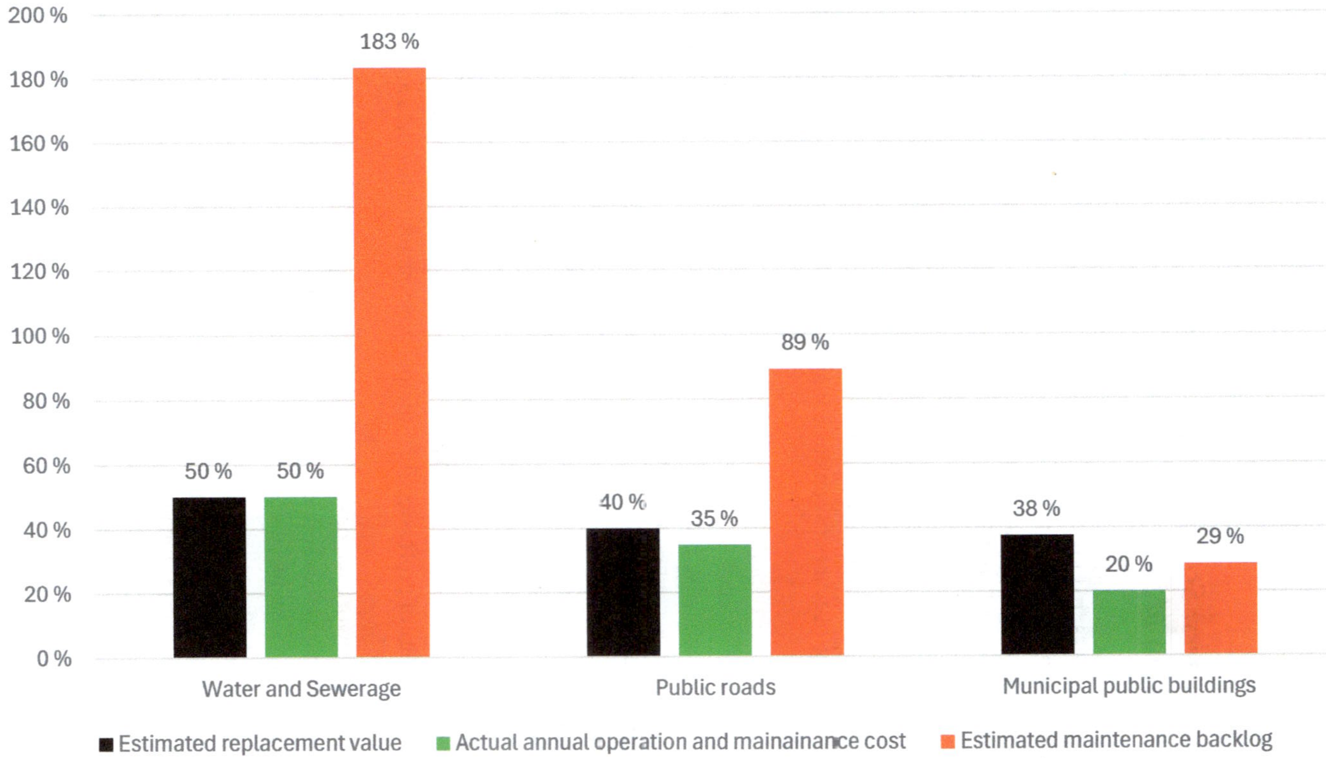

Fig. 6.3 Development in climate vulnerability for physical infrastructure (2010–2025). Estimates of percentage increase measured in constant 2011 NOK for **a** total volume of infrastructure (measured by replacement value), **b** annual operation and maintenance expenditure, and **c** the accumulated maintenance backlog. Compiled by the author based on data from RIF, KOSTRA, Norsk Vann, and the National Transport Plan

This data confirms the central thesis of our research: we are not failing at adaptation because we lack engineering solutions, but because we refuse to stop expanding the very infrastructure that climate change is now threatening to dismantle.

Beyond the Pioneers: Scaling Integration in Sogn Og Fjordane by Means of "Dugnad"

When climate policy was first institutionalized at the national level in Norway, little discussion occurred about creating a unified framework. Instead, a sharp divide emerged: mitigation and adaptation were treated as separate domains, underscored by separate white papers and assigned to different ministries. Initially, the Ministry of Justice held responsibility for adaptation—viewing it through the lens of civil protection—before it was transferred to the Norwegian Environment Agency in 2014. This move prompted strong protests from the Directorate for Civil Protection and Emergency Planning (Direktoratet for samfunnssikkerhet og beredskap [DSB]), highlighting a persistent tug-of-war between environmental and emergency management sectors.

While the state was busy building silos, the local level was busy breaking them down. Following early successes in pioneering municipalities like Fredrikstad, the focus shifted to whether these integrated models could be scaled up to an entire region. Sogn og Fjordane county became the laboratory for this next step. In 2011, the county authority asked WNRI to move beyond merely having an "adaptation chapter" and instead to contribute in developing a fully integrated climate policy.

We organized this effort as a *dugnad*. This uniquely Norwegian term, derived from Old Norse *dugnaðr* ("help, strength"), describes a collective effort where the benefit of the joint outcome exceeds the individual labor investment. In this context, it described a highly professional knowledge-collaboration between researchers, regional authorities and local planners. The goal was to normalize climate work—making it a standard part of professional municipal governance rather than an extra task for specialists.

The analytical framework we developed for Sogn og Fjordane was ahead of its time, anticipating the now-standard IPCC risk terminology by nearly a decade. As shown in Table 6.1, we introduced three critical dimensions:

- *Physical risk* versus *transition risk*: We distinguished between the consequences of a changing climate (natural vulnerability) and the consequences of changing policies and technologies (climate policy vulnerability). Long before transition risk became a global buzzword, we were asking municipalities to analyze how stricter national emission regulations might affect their local industry and transport sectors.
- *The combined effect*: We emphasized that risk is the combination of the climate hazard (e.g. a storm) and the socioeconomic vulnerability (e.g. the local capacity to respond).

Table 6.1 Elements included in our proposal for an integrated climate policy plan[31]

Analysis step	Natural vulnerability (cf. hazard)	Socioeconomic vulnerability (cf. exposure and vulnerability)	Climate policy vulnerability (cf. transition risk)
Current conditions	Impact from current climate and hazards	Exposure and capacity to manage today's challenges	Exposure to existing climate measures and regulations
Future situations	Altered risk due to climate change	Capacity to address tomorrow's climate impacts	Exposure to potential future climate policies and technology
Overall vulnerability	Cumulative effect: How society evolves to become exposed to the combined changes in climate and policy.		
Strategies	Integrated measures: Prioritizing actions that avoid maladaptation and malmitigation.		

- *Avoiding maladaptation and malmitigation*: We systematically analyzed the "clashes" between the two fields—ensuring, for example, that an adaptation measure (like constructing a sea wall) did not accidentally increase GHG emissions (malmitigation), or that a mitigation measure (like urban densification) did not increase pluvial flood risk (maladaptation).

Table 6.1 also illustrates how the framework's original terminology maps against the current IPCC risk framework, in which natural vulnerability corresponds to physical climate risk (specifically hazards); socioeconomic vulnerability corresponds to physical climate risk (specifically exposure and vulnerability); and climate policy vulnerability corresponds to what is now widely termed as transition risk—the societal and economic risks arising from the transition to a low-carbon society.

Everyone Must Set Aside Time for Climate Work—Except the County Governor?

The regional *dugnad*—a traditional Norwegian term for a voluntary collective effort—was structured around three intensive working sessions. Participants from five municipalities—Sogndal, Førde, Eid, Vågsøy, and Jølster—received homework assignments between sessions to apply our integrated framework to their local realities.

While the sessions were full of active discussion, the practical follow-up often faltered. For every representative, climate was just one of many competing responsibilities. At the second session, after several planners admitted they hadn't found the time to prepare, Eline Orheim—from the County Governor's Emergency Preparedness Department, the state's regional oversight body—lost her patience. She closed the meeting with a stern warning:

[31] Dannevig et al. (2014).

Climate is a critical issue. You simply must prioritize your schedules and set aside sufficient time for this work!

Our project plan included a clear "relay" strategy. WNRI would lead the first round of sessions while the regional authorities (the County Municipality and the County Governor) observed from the sidelines. Once they had learned the method, they were to take the "baton" and lead the remaining 21 municipalities in the region on their own.

However, at our final internal project meeting prior to our last session with the municipalities, when I noted that it was now time for the regional bodies to take the lead, Eline Orheim—the same official who had so forcefully lectured the municipalities—sighed and remarked: "But we simply don't have time for this!"

It was a sobering conclusion. The "tyranny of the urgent" had not only caught up with the local planners but had paralyzed the regional leadership as well. The systematic integration of climate policy was not a question of intellectual merit, but of administrative survival.[32]

From Action Research to Critical Analysis: Shifting Focus from "How-To" to "Why-Not?"

From the second half of the 2000s, my focus shifted from conducting research "for" to increasingly conducting research "on" climate change adaptation. This transition was made possible when the Research Council of Norway finally established major programs that included opportunities for social science.

While my earlier work had been characterized by action research—a close, hands-on collaboration with municipalities (often funded by KS) to develop practical "how-to" tools—this new phase allowed for greater analytical distance. We moved from helping systems work to asking why they were not working.

This academic turn was reinforced by the establishment of a long-term Nordic research collaboration. Through two succeeding networks financed by the Nordic Council of Ministers, we initiated a series of working seminars that brought together leading researchers from across the Nordic region. This collaboration was not only a catalyst for our own theoretical development but also led to the establishment of the bi-annual Nordic Conference on Climate Change Adaptation (NOCCA), with the first held in Stockholm in 2010.[33] This regional network proved to be a vital stepping stone, providing the institutional weight and international anchoring that would

[32] Aall (2005b).

[33] Wallgren (2011).

later prove essential for establishing the Norwegian Research Centre for Sustainable Climate Change Adaptation (Noradapt).

Armed with this broader Nordic perspective, we began to investigate the underlying institutional, political, and economic structures across sectors like insurance, agriculture, and energy.

A Theory on How the Climate and Emergency Management Sectors Can Co-create Climate Change Adaptation

In our first major project in this more classical research vein, named "Civil Protection and Climate Vulnerability" (CIVILCLIM), we compared Norway, Sweden, and the Netherlands. We analyzed how the emerging field of adaptation was being fathered by two very different parents: the established, proactive climate mitigation policy and the traditional, reactive emergency preparedness sector.

As shown in Table 6.2, these sectors operate with fundamentally different value systems. Mitigation is civilian-led and highly political, while preparedness is often influenced by military traditions and often remains depoliticized. Our finding at the time of the study was that adaptation had largely been swallowed by the preparedness logic. It remained "low politics"—technical and reactive—aimed at protecting society primarily from extreme weather events rather than transforming it to be less vulnerable.

However, as the field of climate change adaptation matured, it gradually became more influenced by the climate policy domain. This shift was underscored when the national coordination responsibility for adaptation in Norway was eventually transferred from the DSB to the Norwegian Environment Agency in 2013.

Are Natural Hazard Events Caused by Severe Weather or Poor Planning?

As we moved further into critical analysis, it became a mantra for me to ask: *Is it climate change or societal change that contributes most to climate risk?* In the project Land-Use Planning and Preparedness for Future Climate (AREALKLIM), led by my colleague Halvor Dannevig, we addressed this by analyzing ten historical natural hazard events[34] alongside five ongoing municipal planning processes,[35] all of which were taking place in Western Norway.

Our *first* conclusion was blunt: poor planning was the primary cause of the natural hazards we studied. Out of ten events—two of which involved loss of life—only one could reasonably be linked to climate change: a landslide at Hatlestad Terrace in

[34] Dannevig et al. (2013).

[35] Dannevig et al. (2015).

Table 6.2 Characteristics of existing institutions for climate policy versus civil protection that may influence the emergence of climate change adaptation as a distinct policy area[36]

Institutional characteristics	Climate policy domain	Climate change adaptation domain (emerging)	Civil protection and preparedness domain
Level of political conflict	High	→?←	Low
Emphasis on policy measures	Proactive	→?←	Reactive
Administrative tradition	Civilian	→?←	Military
Administrative capacity	High	→?←	Low
Main role of NGOs in policymaking	Agenda setting	→?←	Implementation
Climate policy focus	Mitigation	→?←	Adaptation
Dominant developmental paradigm	Transformative	→?←	Protective
Scenario selection bias	Mid-range alternative	→?←	Worst case
Disciplinary tradition in risk analysis	Natural sciences	→?←	Social sciences
Emphasis on climate change events	Slow onset	→?←	Extreme
Perception of climate change causality	Anthropogenic	→?←	Indifferent
Weight of climate vs. societal change as risk driver	Climate change	→?←	Equal

Bergen that killed three people. In that specific case, the precipitation was so extreme it far exceeded what anyone could have anticipated when the land-use plan was established in 1967. For the other nine cases, the disasters were essentially human-created through a failure to respect known risks or maintain existing infrastructure.

Our *second* conclusion was equally sobering: the Norwegian planning system is built for yesterday's climate. We found that if spatial planning is conducted strictly according to current legislation, it ensures adequate protection only as long as current climate conditions persist. However, the system is poorly equipped to handle a "wilder and wetter" future. We identified several systemic weaknesses:

- *Weak national policy signals*: A lack of clear, mandatory requirements for climate change adaptation.
- *Capacity gaps*: Small and medium-sized municipalities simply lack the resources for complex hazard assessments.

[36] Groven et al. (2012).

- *Depoliticization*: National "simplification" reforms have reduced the scope of state objections, allowing local development interests and market forces to override environmental and safety concerns.

Climate change adaptation risks remaining in the category of "low politics"—a technical side-issue that must yield whenever it conflicts with powerful economic interests or local growth ambitions.

The Leikanger Calendar: An Archaeological Warning

One of our most profound insights in the AREALKLIM project came not from a spreadsheet but from a local history calendar. During our analysis of the planning process in Leikanger (now part of Sogndal municipality), my colleague Lisbeth Dahle and I were struck by a photograph in the 2014 "Leikanger Calendar." It showed an archaeological trench where the soil layers told a dramatic story of prehistoric disaster. The image revealed a thick band of landslide deposits sandwiched between two dark soil layers—one from 1400 BCE and the other from 500 BCE. I contacted the archaeologist Asle Bruen Olsen from the University of Bergen, who was responsible for the excavations. He explained that this pattern was consistent across sixty similar excavations in Western Norway. He also presented his hypothesis that the cause of this was a "double hit" of environmental and societal change: while a transition to a wetter, colder climate had occurred, it was the human response—extensive deforestation to free up grazing land for a growing population—that had destabilized the slopes. By removing the forest cover, the prehistoric inhabitants had inadvertently stripped the landscape of its natural protection.

This discovery became a powerful analytical tool for us in two ways:

- *Challenging the sceptics:* Within our own project team, the Quaternary geologist Rune Aa—a skilled scientist but a noted climate sceptic—had argued that since landslides were frequent during the cold Bronze Age, a warmer climate would not necessarily increase landslide frequencies today. However, the Leikanger data changed the conversation. It proved that temperature is only one variable; land management, the integrity of the forest cover, and the root systems that bind the soil are the others.
- *The message to planners:* We used this "prehistoric warning" to demonstrate to Leikanger municipality that their current landslide risk was not merely an "act of God." It was a direct consequence of how forest removal leaves steep terrain vulnerable to the elements.

The lesson for AREALKLIM was clear: climate change does not act in isolation. It is the combination of increased precipitation and human-driven deforestation that creates catastrophe. Managing the global climate is impossible for a municipality, but managing the local landscape—specifically through the strict protection of forest cover in hazard-prone areas—is a choice they have the power to make.

The Battle of the Spruce: When Climate Change Mitigation Becomes a Risk

The story of spruce planting in Western Norway is a textbook case of how climate measures in one area can create dangerous risks in another. Following World War II, Norway embarked on a massive national project to "build the country" by planting non-native Norway spruce across the west coast, aiming to make the nation a leader in paper production. Decades later, this was rebranded as a climate mitigation measure, sequestering carbon in fast-growing "industrial forests."

However, our research in a project named "The Impact of Increasing Spruce Plantation Area on the Carbon Balance of Forests in Western Norway" revealed two inconvenient truths that challenged both the forestry industry and national climate policy.[37]

First, the carbon benefit is highly conditional. Our analysis showed that the climate impact of these forests depends more on whether the wood substitutes for carbon-intensive materials like steel or concrete, rather than the tree species itself. If the timber is instead used for luxury holiday cabins, the net climate benefit disappears or even becomes negative.

Second, and more critically for adaptation, we found a significant increase in physical climate risks. Data compiled during our research showed that nearly one-third of the spruce planted in Western Norway is located on terrain steeper than the 27° landslide hazard threshold, as defined by the Norwegian Water Resources and Energy Directorate.[38] As climate change brings more intense rainfall, these heavy, shallow-rooted monocultures on steep slopes significantly increase the likelihood of landslides.

When Interests Trump Insights—The "Carlo Aall Spruce Campaign"

In the autumn of 2013, landslides battered Western Norway. My neighbor and Chief Engineer at the Norwegian Public Roads Administration, Nils Magne Slinde, estimated that repairing a single one of these incidents would cost 30 million NOK. When the national broadcaster (NRK) interviewed me two days later, the journalist clearly expected me to blame global warming.

[37] Simonsen et al. (2022).

[38] Norges vassdrags- og energidirektorat (NVE) (2019).

I didn't. Instead, I pointed to the massive spruce plantations that have carpeted Western Norway over the last 50 years—often on dangerously steep slopes. This was the result of a post-WWII national strategy to "build the country" by making Norway a world leader in paper production. We essentially terraformed the coast with non-native spruce.

Shortly afterwards, a colleague working in Sogn og Fjordane County municipality warned me about a targeted campaign initiated by Coastal Forestry (*Kystskogbruket*), an interest group representing regional authorities and the timber industry. Their manager had sent emails to stakeholders along the Norwegian coast from Rogaland in the south to Nordland in the north, claiming I was "discredited as a professional" because of my political background in the Socialist Left Party (at the time I was a member of that party). She argued that my warnings against spruce planting were a "disguised political agenda" to promote nature conservation at the expense of the forest industry, and that there was "no scientific basis" for claiming that these forests increased landslide risk.

I found it more amusing than threatening. A year later, I even invited the same organization to join a research application to study exactly this: the link between spruce planting and climate-related landslide risk. Their response was telling:

> We find the topic relevant and would, in principle, like to join the user group, However, we note that Western Norway Research Institute is listed as a partner with Carlo All, covering the fields of climate policy, vulnerability, and adaptation. We believe other research environments would cover this field better.

Naturally, they didn't join. More importantly, the Research Council of Norway rejected the application—both that year and again two years later. This was a stark lesson in the barriers to systemic change: when research challenges the economic and political foundations of an established sector, "knowledge co-production" often stops, and "gatekeeping" begins.

Insurance Dead End: Security Bureaucracy Versus Climate Transparency

In 2013, we initiated a groundbreaking collaboration with Finance Norway (the national industry organization for the Norwegian financial sector) and Jan Ketil Rød at the Norwegian University of Science and Technology (NTNU). The premise was simple but powerful: for years, the insurance industry had sat on a goldmine of insurance loss data—specifically data on payouts for various natural hazards, such as water ingress in basements, urban flooding, and storm damage. While the industry had detailed records of the escalating costs (see Fig. 6.4), municipalities were largely conducting spatial planning and infrastructure management in the dark, relying on fragmented public records.

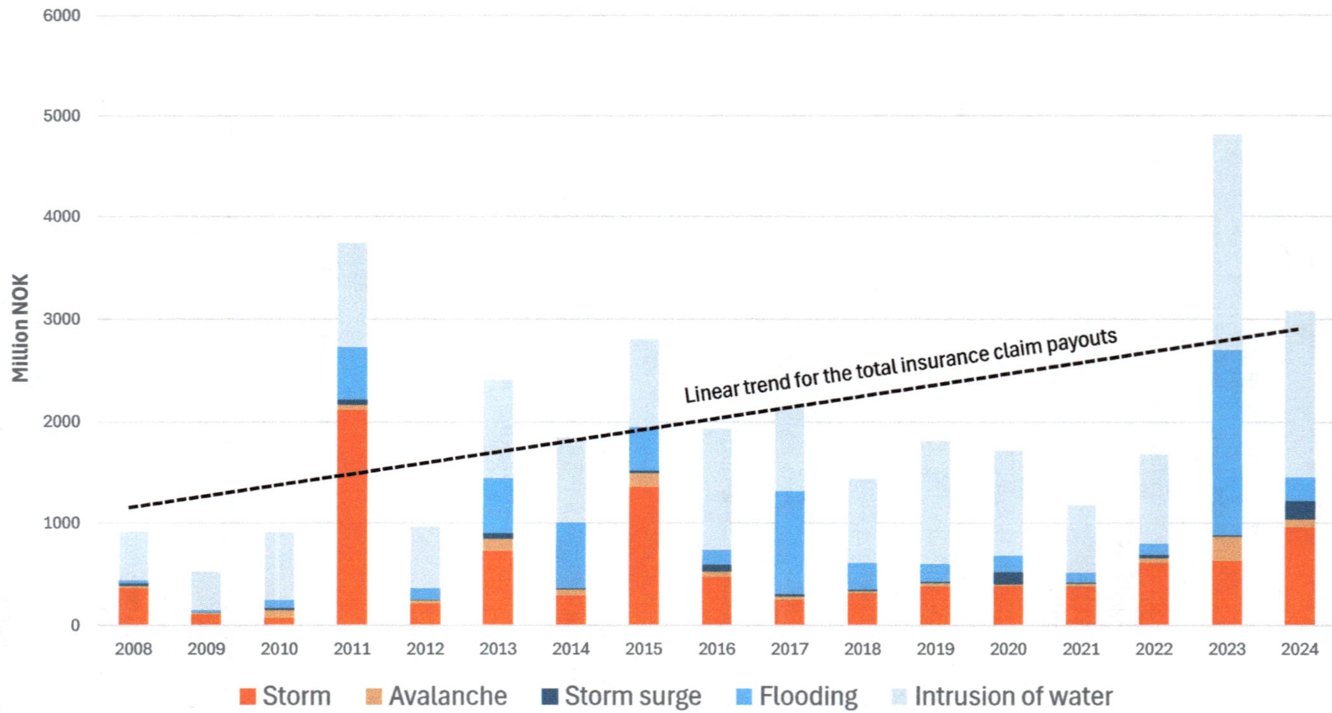

Fig. 6.4 Development in total payouts for climate-related natural and water damage for all of Norway (constant 2024 NOK, 2008–2024). The linear trend highlights a significant increase in claims, particularly driven by "intrusion of water" and "flooding"

We conducted a pilot project with nine municipalities, including Bærum, Stavanger, and Trondheim. We provided them with access to payout data, geolocated down to the address level of individual households. The conclusion was unanimous: this data proved incredibly valuable. It allowed planners to see exactly where the "shoe was pinching"—identifying recurring damage patterns, such as chronic sewer backups or localized surface water issues, that were invisible in standard topographical maps or historical disaster records. We recommended a national scheme where all municipalities could access these high-resolution loss data.[39]

At first, the follow-up showed promise. In 2017, Finance Norway, the DSB and the Norwegian Mapping Authority began planning a national online service. In 2019, the website Kunnskapsbanken (The Knowledge Bank) launched with great fanfare. For a brief period, municipal employees had exactly the tool we had tested: georeferenced claims data down to the local level.

The enthusiasm, however, was short-lived. Only months after the launch, access to the data was pulled. Citing the Security Act and the threat of terrorism, national security authorities determined that such detailed aggregation of vulnerability data could expose critical infrastructure to sabotage. When the service was later reopened, it was in a significantly reduced version. Data were available only at a coarse municipal level and had up to 2025 not been updated since 2018. Furthermore, the process became heavily bureaucratized; municipalities now must apply manually for access through the County Governor. Finance Norway has expressed strong frustration over this "institutional dead end," noting that a potentially dynamic planning tool has been stifled by a security bureaucracy that prioritizes secrecy over climate transparency.

However, in early 2026, a long-awaited breakthrough finally occurred. After years of persistent pressure from the research community and the insurance industry, DSB has finally moved to update the Knowledge Bank with fresh claims data.[40] While the battle between security and transparency continues, this reopening marks a vital shift: an acknowledgement that the tangible, escalating risk of keeping municipalities in the dark about their own climate vulnerability is far greater than the hypothetical risk of terrorism.

Today, the Knowledge Bank stands as a sobering, yet ultimately hopeful, example of how persistent advocacy can eventually restore the transparency needed for societal survival in a changing climate.

Western Norway Research Institute's Homemade Vulnerability Model Meets the IPCC Risk Framework

The summer of 2018 marked a turning point for our analytical approach. During a family holiday in Provence, I met my long-time colleague Ghislain Dubois for a

[39] Breivik et al. (2014).

[40] DSB (2026).

working lunch in Marseille. Overlooking the harbor, we sketched out the foundation for what would become the UNCHAIN project ("Unpacking Climate Impact Chains").

Ghislain Dubois, now representing the international consultancy Ramboll, wanted to draw on a methodology developed by his close friend and colleague, Marc Zebisch, at Eurac Research. Based in Bolzano, Italy, Eurac is a private research center renowned for its interdisciplinary work on complex societal challenges, particularly in mountain regions. Their "Impact Chain" methodology was exactly what we needed.[41] It involves tracing cause-effect chains backwards—from an assumed risk to the specific intervention points where adaptation is most effective.

For us at WNRI, this was a major leap. We were moving from our relatively simple, localized vulnerability assessments to the IPCC's latest, sophisticated framework for climate risk analysis, anchored in the expertise of some of Europe's leading risk researchers.

The UNCHAIN project, which I co-led with Ghislain, became a massive undertaking involving nine partners from six countries. We tested the impact chain methodology across 11 case studies, covering everything from tourism and transport to transboundary risks.[42] The project resulted in a record 14 scientific publications, culminating in a dedicated Research Topic featured by *Frontiers*.[43]

For me personally, the most rewarding intellectual work within UNCHAIN was developing a new conceptual road map for how to describe and address the inherent uncertainties of climate change adaptation. In an article co-authored with my colleague Kyrre Groven—drawing on longitudinal studies from Voss municipality[44]—we challenged the naïve tendency in climate policy to treat climate model outputs as objective, programmable facts.[45]

Our core message was a critique of the traditional "predict-then-act" model. We argued that wasting resources on attempting to predict the exact probability and location of future events is often a futile exercise, given the immense complexity of both climate and societal systems. Instead, we proposed a "reflect-then-act" approach. This framework consists of three transformative shifts:

- *Accepting irreducible uncertainty*: We must move away from the illusion that more research will always reduce uncertainty. In many cases, we simply will not be able to "quantify it away," and our planning systems must learn to function within that reality.
- *From technical to multidimensional uncertainty*: Instead of viewing uncertainty as a one-dimensional technical problem (the "error margin" in a climate model), we must address it as a multidimensional challenge. This includes acknowledging

[41] Fritzsche et al. (2014).

[42] Aall (Ed.) (2023b).

[43] Aall et al. (2023c).

[44] Dannevig et al. (2016).

[45] Aall and Groven (2022).

the deep uncertainties embedded in social and political processes and the systems we are trying to protect.

- *Applying the precautionary principle*: We argued for a shift in how the precautionary principle is used. While it is a staple in climate change *mitigation*, it is often overlooked in climate change *adaptation*. By shifting to a "reflect-then-act" model, we accept that we must act even under high and varied forms of uncertainty, prioritizing resilience over precise prediction.

This approach acknowledges that managing the local landscape is not a mathematical puzzle to be solved, but a continuous process of social and political reflection.

UNCHAIN also allowed my colleagues at WNRI and me to explore the (for us) new concept "transboundary climate risks". In the Norwegian component of a multi-case study, we analyzed how representatives of local agriculture and the municipal administration in a municipality dominated by intensive livestock farming reacted to this new layer of complexity. For these stakeholders, climate risk was no longer just a question of how intensified rainfall might hamper local harvesting; it was also about how climate shocks elsewhere—such as droughts in Brazil—could disrupt global supply chains, affecting the import of soy for concentrated cattle feed and threatening local economic stability.[46]

This international collaboration provided us with more than just new tools; it validated our long-standing focus on societal vulnerability. By integrating our "home-made" insights into a rigorous international framework, we were now ready to lead the national conversation on a much larger scale.

The Unthinkable Idea: Reducing Mobility to Mitigate Climate Risk

In 2021, the Office of the Auditor General of Norway issued a serious criticism regarding the state's failure to ensure an adequate overview of climate change impacts on physical infrastructure.[47] This is the second-highest out of four levels of formal reprimand in the Norwegian system. The investigation pointed out that the Norwegian Public Roads Administration (NPRA) had consistently assessed risk based on current, rather than future, climate conditions.

Against this backdrop, I was commissioned in 2022 to lead the Climate Road (*Klimavei*) project, together with the Norwegian Geotechnical Institute (NGI) and Menon Economics, for the NPRA and Nye Veier AS (a government-owned company established to accelerate the development of major highways). Our goal was to

---

[46] Aall et al. (2023a).

[47] Riksrevisjonen (2021).

develop a new methodology for analyzing physical climate risk and for assessing the socioeconomic costs and benefits of adaptation measures.[48]

We tested our approach on five specific road sections, building on the *Impact Chain* framework from the UNCHAIN project. We also revisited a model of local climate vulnerability that we had developed in 2008, adapting its focus from municipalities to roads.[49] This led us to identify *five key drivers of climate vulnerability* for transport corridors:

- *Mobility*: Projected changes in the volume and type of transport.
- *Population*: Expected changes in the size and age structure of the population within the functional regions connected by the road (e.g., the impact of hospital closures or the centralization of dairy production, which increases the necessity of transit).
- *Physical infrastructure*: The technical condition of the road and the ecological status of the surrounding watershed.
- *Economic development*: Trends in local industry, such as forestry or agriculture, within the watershed.
- *Mentality*: Societal attitudes toward mobility and risk—a factor that often proves to be the most stubborn barrier to change.

Our goal was to provide road planners with a broader set of strategies than the typical "repair or protect" approach. We structured these strategies within the core components of the IPCC risk framework (hazards, vulnerability, exposure, response, and risk) to show exactly how different interventions (response) target the remaining four components (see Fig. 6.5):

- *Reactive* adaptation measures minimize consequences after a *risk* materializes. They are designed to manage outcomes, such as damage and injuries, from unavoidable events. Example: Enhancing the capacity for rapid reconstruction and reopening of a transport corridor.
- *Protective* adaptation measures address *hazards.* They aim to block or divert a climate hazard (e.g., an expected increase in precipitation) before it impacts the infrastructure. Examples: Constructing landslide sheds, avalanche barriers, or flood walls.
- *Proactive* adaptation measures address *vulnerabilities.* They reduce the inherent susceptibility of the road system, making it more resilient when an impact occurs. Examples: Addressing the maintenance backlog, upsizing culverts, or improving drainage.
- *Reducing risk-acceptance levels* addresses *exposures.* This involves a deliberate decision to lower the threshold for acceptable negative impacts (e.g., lowering the threshold for closing a road). Examples: Adopting stricter safety standards, accepting a higher frequency of road closures, or reducing speed limits during extreme weather.

[48] Aall et al. (2025a).

[49] Holm and Aall (2022).

- *Transformative* adaptation measures address the *drivers of societal change.* They lower vulnerabilities by addressing the societal trends that create reliance on high-risk corridors. Example: Implementing land-use planning that decreases transportation demand, thereby reducing society's dependence on a specific transport corridor.

For completeness, the *mitigation measures* included in the figure address the drivers of climate change by reducing GHG emissions (e.g., transitions to electric vehicles). While essential for preventing future hazards, mitigation fell outside the immediate scope of this specific adaptation project.

While the road authorities were receptive to this framework, we encountered significant resistance from the economist partners in the project. In practice, our attempt to operationalize vulnerability was stifled by the economic assumption of perpetual growth in transport volume. By keeping mobility growth as an immutable given in their risk assessments, the economists effectively concluded that physical "protective measures" were the only viable solution. Varying or reducing transport volume—a key driver of risk—remained a non-option in their models.

"We Drive Straight Ahead, No Matter What We Hit"

To understand how the mentality factor operates in practice, I often recall a playground game from my childhood. We would link arms in a long row and march forward, singing: "*Vi går rett frem, samme hvem vi går på*"—which translates to: "We walk straight ahead, no matter who we run into." The goal was unwavering forward momentum; anyone in our path simply had to get out of the way or be pushed aside.

This childhood logic—a mix of tunnel vision and the refusal to deviate from a set path—perfectly illustrates our modern relationship with transportation and climate risk. I see this play out on my own commute between Leikanger and Sogndal. The road, notorious for landslides since it opened in 1937, finally began to be secured with tunnels in 2007.

Tragically, six months into construction, the very accident the tunnels were meant to prevent occurred: a young teacher—who taught one of my own children—died from a rockfall at Fatlaberget. The community's immediate reaction was one of collective caution. 850 people signed a petition demanding an alternative boat service. For a while, the boat became a symbol of shared safety.

But a strange thing happened. As the weeks passed, usage dwindled. Soon, I was one of the few still taking the boat; everyone else had returned to their cars and buses, driving right under the dangerous mountainside again.

When this road opened in 1937, I was told that no one would dream of driving in bad weather—the risk was common knowledge. Today, we have built ourselves into a technological expectation of 24/7 availability. We have reached a state of "unwavering forward momentum," where we expect to drive

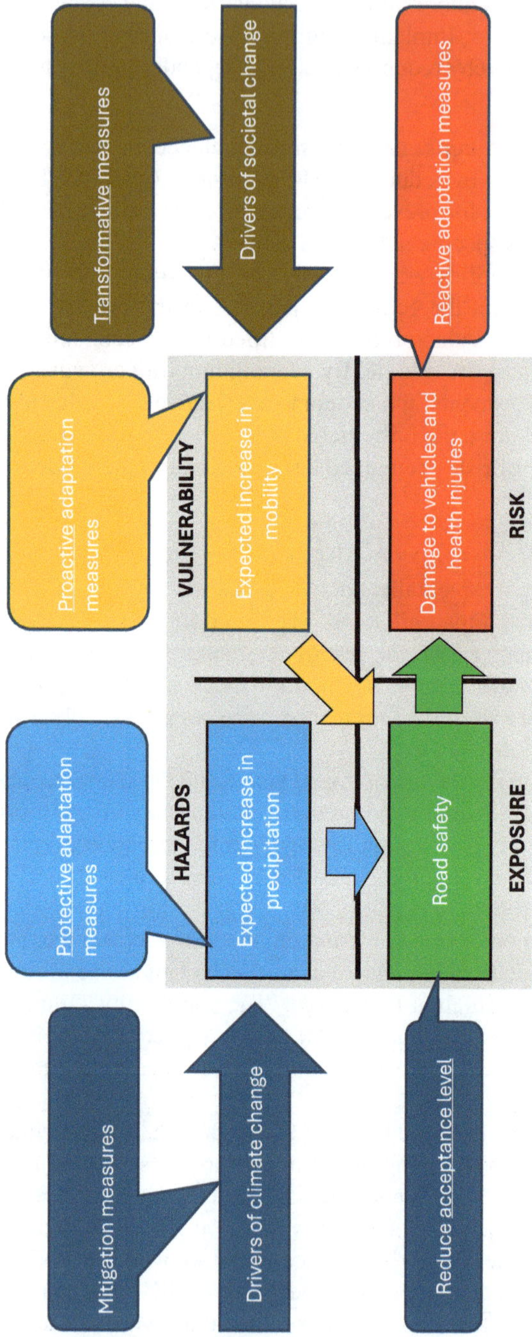

Fig. 6.5 Strategies addressing various components of the IPCC framework for analyzing physical climate risk

straight ahead regardless of the weather or the landscape. This mentality is perhaps our greatest vulnerability: it is the psychological "impact chain" that locks us into expensive, high-risk infrastructure rather than allowing us to consider staying home or taking the boat.

The Blind Spot: When the Solution Becomes the Risk

While the research community has spent decades discussing *maladaptation*—where adaptation measures backfire or shift vulnerability—we have almost entirely ignored its mirror image: *malmitigation.* In the ongoing project Creating Sustainable Renewable Energy Futures with Low Climate Risks (SUSRENEW), together with colleagues from Aalborg University, NORCE, and IFE, we are asking a question that some find almost inappropriate: What if our primary solution to the climate crisis—the transition to a 100% renewable energy system—increases our physical climate risk?

A search of the *IPCC Sixth Assessment Report* tells the story of this blind spot: Maladaptation appears 581 times, while maladaptation appears only twice. To understand why this matters, we must look at the tension between global goals and local realities. In the SUSRENEW project, we are developing a theoretical framework to understand if, and eventually how, we are building ourselves into a systemic trap. We are testing three hypotheses on how our current path might lead to a "wicked feedback loop":

- *The growth trap*: The pursuit of continued economic growth, even when coupled with climate targets, drives a massive expansion of physical infrastructure for energy production and distribution. This infrastructure is itself exposed to climate hazards. Furthermore, renewable energy currently supplements rather than replaces fossil fuels, thus increasing the total volume of vulnerable assets in our society. Since 1990, the global share of primary energy consumption from renewables has only increased from 6.8 to 14.8%. This is because total energy consumption has surged by 84% during the same period, while the absolute volume of fossil fuel use has also increased by 71%.[50]
- *The complexity trap*: Drawing on the work of sociologist Charles Perrow,[51] we argue that as our energy systems become more tightly coupled and complex (through digitalization and decentralization), the likelihood of large-scale system failures increases when faced with external shocks like extreme weather events.[52]

---

[50] Our World in Data (2026a).

[51] Perrow (2007).

[52] Aall et al. (2022).

- *The electrification trap*: The vision of electricity powered by "weather" as a "universal energy carrier" creates a single point of failure. Developing a system capable of supplying electricity for all purposes requires a massive grid expansion, making society extremely fragile if compound events—such as a simultaneous drought in Norway and a cold snap in Europe—disrupt the supply.[53]

If these hypotheses hold true, we face a phenomenon where increased GHG emissions and heightened physical climate risk reinforce each other in a negative spiral. The climate issue has long been described as a super wicked problem, two of its primary features being that time is running out, and those tasked with solving it are also those causing it.[54] The ongoing experiment of massive renewable energy expansion—justified by the climate goal of a net-zero society, yet lacking a mechanism to ensure that new renewables replace rather than merely supplement existing fossil fuels—may introduce a new "super wicked" mechanism: a phenomenon we have tentatively termed the "wicked feedback loop." We aim to address the hypothesis that the very pursuit of a net-zero society, if driven by the same logic of growth and complexity that created the crisis, may lead us straight into a systemic vulnerability trap.

"No Wind Turbines Shall Be Built Here"

The conflict between climate mitigation and climate risk is not just a matter for academic journals; it is a deeply felt cultural tension. I've seen this firsthand through my hobby as a musician. I play the bass in a band with the distinctly local name *Grev Moltke og Dei Raude Prinsane* (named after traditional pear and apple varieties from Sogn).

My good friend and former national park manager, Tom Dybwad, wrote a song for us that alludes to the issue of malmitigation by playing on the strong opposition to onshore wind power in Norway. While the song is a protest against the eternal scars left by wind farm roads, it captures a core dilemma: you cannot "save the climate" by industrializing the very nature that provides our resilience. The chorus goes like this:

> There shall be no wind turbines built here,
> And there shall be no wind turbines built there.
> There shall be no wind turbines built neither there nor here.
> So then, there will be no wind turbines here.

You can check it out on Spotify (search "Prinsane" and "Vindturbinsongen").

---

53 Aall et al. (2025b).

54 Levin et al. (2012).

Establishment of the World's First Research Centre for Sustainable Climate Change Adaptation (Noradapt)

An Academic Community of Practice: The Birth of Noradapt

The establishment of the Norwegian Research Centre for Sustainable Climate Change Adaptation (Noradapt) in 2018 was not merely the creation of another research institute. Formally, Noradapt is organized as a project, albeit a large-scale and long-term one. It operates with funding granted for one year at a time, which means it has no fixed budgetary endpoint but instead requires annual renewal through the national budget. This organizational structure represented a strategic effort to bridge the gap between fragmented local knowledge and the systemic national challenges of climate change. The center began as a collaborative effort among eight—and later nine—research institutions across Norway, including partners such as CICERO, NORCE, Nordland Research Institute, the University of Bergen, NTNU, SINTEF Community and the Western Norway University of Applied Sciences, with WNRI as the lead.

The timing of Noradapt's founding was a result of a specific political window of opportunity. At the time, Norway had a minority government where the Liberal Party (*Venstre*) was a coalition partner. Dependent on the Christian Democratic Party (*Kristelig Folkeparti* [KrF]) for budget approval, the government remained open to negotiations. Through our contacts in the Liberal Party—specifically representatives from our own county who were part of the government apparatus—we received strategic advice to approach KrF's representatives. We realized that if KrF included Noradapt in its list of demands for supporting the national budget, the project would be secured. This political maneuvering worked, largely because WNRI had already gained recognition as a national leader in the field. Furthermore, WNRI's location in Sogndal—a region highly exposed to climate-related hazards—provided a distinct advantage. This proximity increased the legitimacy of our work, allowing the region to function as a proximate laboratory for studying natural risks and adaptation strategies—a grounding often missing from urban boardrooms.

One of our first major tasks was to operationalize the concept of sustainable climate change adaptation. In doing so, we highlighted a profound imbalance in how society manages climate risk. As illustrated in Fig. 6.6, a massive discrepancy exists between what society is willing to pay to repair damage and what it is willing to invest in preventing it through knowledge production. While billions are spent annually on restoring infrastructure after floods and landslides, only a tiny fraction is allocated to researching how to mitigate these impacts. The figure compares the massive insurance payouts for water and natural hazard damage in 2018 (over 4100 million NOK) against the disproportionately small investment in climate change adaptation research (13 million NOK). This gap highlights the systemic underfunding that Noradapt was established to address.

Furthermore, we identified a significant academic bias within the limited funds set aside for preventive knowledge. Most research funding was—and still is—directed toward natural sciences and technology, leaving social science-based adaptation

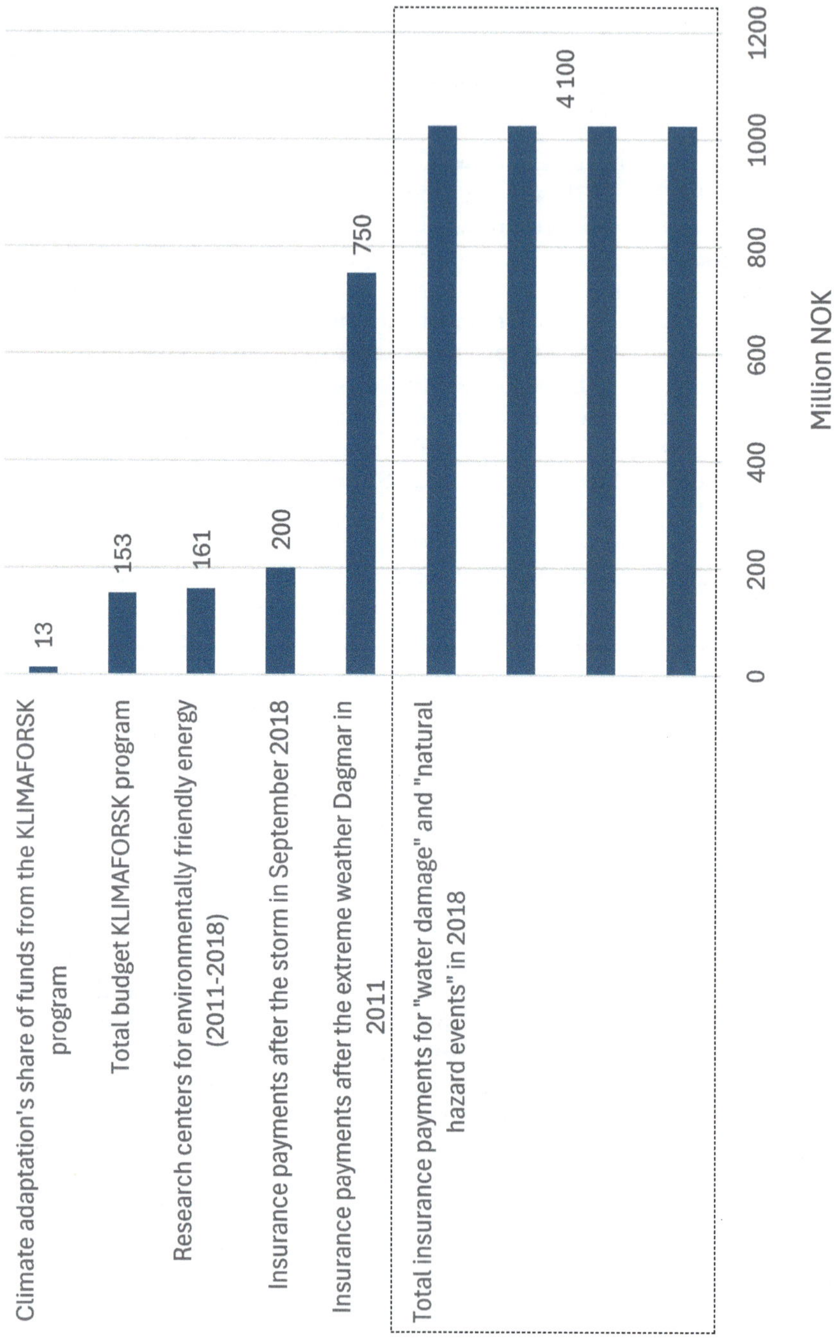

Fig. 6.6 Funds allocated to various research purposes and to cover damages from climate-related events

research critically overshadowed. This lack of a social science perspective constituted a major blind spot; climate change is as much a societal challenge as a physical one. Our concern was that as climate impacts become more evident, autonomous adaptation will occur regardless of government policy. Without a solid social research-based knowledge base, the risk of negative side effects increases. We saw a pressing need to ensure that adaptation measures do not conflict with other sustainability goals, such as social equality or emission reductions.

Noradapt's role was to teach the various sectors to use a systemic framework, similar to the one applied in the road transport sector (see Fig. 6.5)—to identify where interventions target hazards, vulnerability, or the drivers of societal change. By defining the boundaries for what constitutes sustainable adaptation, we aimed to prevent maladaptation from becoming the default response to a changing climate. Furthermore, the establishment of Noradapt allowed us to formalize our position at the forefront of Norwegian climate research, transitioning from competing for individual projects to leading a long-term national effort.

Since the opening in December 2018, I have presented Noradapt to a wide range of government and parliamentary politicians. These encounters have often revealed how the message of adaptation can be interpreted through different political lenses. On 11 March 2022, WNRI was visited by the then-Minister for Research and Higher Education, Ola Borten Moe. I delivered my usual presentation, highlighting the low priority given to adaptation compared to emission reduction. The Minister took note of this point, but in a way I had not entirely anticipated. After the presentation, he commented[55]:

> I believe it is not effective climate policy for Norway to spend such large resources on reducing domestic greenhouse gas emissions; but when it comes to climate change adaptation, I think it is important that Norway increase its national efforts.

This encounter illustrated a core tension in our work. While we argued for the necessity of adaptation as a *supplement* to mitigation, it was heard by some as a potential *substitute*. It seems there is some rationale behind the environmental movement's long-standing skepticism toward emphasizing climate change adaptation efforts. If adaptation is used as an excuse to delay emission reductions, we risk falling into the very wicked feedback loops we aim to avoid.

Bridging the Indicator Gap: How Noradapt Succeeded where the State Gave up

One of the most persistent challenges in global climate policy is measuring progress. While mitigation has a universal metric—tons of CO_2-equivalents—adaptation lacks a comparable unit of measurement. How do you quantify a disaster that did not happen? This difficulty became clear when Norwegian authorities attempted

[55] Moe (2022, March 11).

to establish national indicators for climate change adaptation. Despite several well-intentioned efforts, they largely abandoned these attempts in favour of more qualitative assessments.

The failure was not primarily due to a lack of data, but rather the sheer complexity of defining and communicating adaptation results. As my colleague Torbjørn Selseng has explored in his doctoral thesis, measuring the effect of adaptation measures is an almost impossible task.[56] Unlike emission reductions, where a saved ton is a saved ton, adaptation is riddled with "ifs" and "buts". Is a reduction in flood damage due to a new levee, or simply because it didn't rain as much this year?

The Norwegian Environment Agency has openly acknowledged that establishing indicators for the *effect* of adaptation is too complex to communicate effectively to the public and decision-makers. The agency has explicitly stated that no methodology currently exists to determine whether society's vulnerability to climate change is decreasing or increasing.[57] Consequently, authorities have been forced to rely on "proxy indicators"—counting administrative outputs, such as the number of municipalities that have conducted risk analyses or incorporated adaptation into their master plans.[58]

This vacuum of meaningful, impact-based information became a crucial backdrop for Noradapt's mission. Where the state's indicator projects had reached their limit, we saw an opportunity to provide a different kind of analytical framework. The inspiration for a new approach came during a study trip to Arendalsuka—Norway's annual democracy festival. Together with Eivind Brendehaug, administrative coordinator at Noradapt, I attended a launch by the NGO Sabima, which had developed an ingenious indicator-based system for ranking municipalities according to their efforts in preserving biodiversity. We decided that Noradapt should achieve something similar for climate change adaptation—effectively stepping in to solve the methodological puzzle that the national authorities had set aside. Back in Sogndal, I established a project team including Professor Jan Ketil Rød (NTNU) and Torbjørn Selseng to develop what would become the world's first system for ranking a country's municipalities based on physical climate risk.

The Importance of Understanding Climate Vulnerability and Why Climate Sceptics May Have a Point

A few months before we presented our system for ranking Norwegian municipalities according to physical climate risk, I wrote an op-ed that generated considerable media attention.[59] The summer of 2023 was marked by large

[56] Selseng (2026).

[57] Miljødirektoratet (2026).

[58] Selseng et al. (2021).

[59] Aall (2023a, August 7).

forest fires in Southern Europe. Many people linked these fires exclusively to climate change. However, during a study trip to Portugal, a local forest ranger explained the real trigger: replacing native pine with eucalyptus—a highly flammable species.

I deliberately chose the somewhat provocative title "Climate Sceptics Are a Little Bit Right," arguing that societal changes are often a more important risk-driver than climate change. My point was simple: if we always blame the climate, we may prescribe the wrong responses. In this case, "fewer eucalyptus trees" matters more than "more firefighters."

The op-ed generated several positive emails from self-declared climate sceptics. This time, the mudslinging came from climate advocates, including this:

> You clearly do not understand that if you give the "climate sceptics" an inch ("a little credit"), they will take a mile.

While I received pushback from climate advocates who feared giving any credit to sceptics, it remains a fundamental point in a polarized world that without recognizing societal vulnerability, we will fail in our adaptation efforts.

Our ranking system, presented in 2023,[60] aligns with the IPCC framework by categorizing risk drivers into *climate hazard*, *vulnerability*, *exposure*, and *response*. The methodology is based on indexing and aggregating multiple indicators from diverse data sources, effectively summing up apples and oranges to create a comprehensive risk profile.

The indicator list presented in Table 6.3 is from the 2024 Norwegian version of the system. This version represents a significant expansion of the initial 2023 edition (from which the map in Fig. 6.7 is derived, and which was also produced in an English version). While the 2023 version focused primarily on vulnerability, the 2024 update incorporated indicators for exposure by including historical insurance claim data from Finance Norway. This addition, as discussed in the previous chapter on insurance data, allowed for a much more robust calculation of physical climate risk.

Using these indicators, we classify all Norwegian municipalities into four risk groups based on the relationship between their climate impact and their adaptation response:

- *Green*: Municipalities that have made greater efforts than others to adapt to climate change, while simultaneously being less affected.
- *Yellow*: Municipalities that have made greater efforts than others to adapt but remain more affected than others due to their location or exposure.
- *Grey*: Municipalities that have made less effort to adapt but are also less affected than other municipalities.

[60] Klimamonitor (2026c).

Table 6.3 Indicators used in the Noradapt system for risk-ranking of municipalities

Risk driver	Indicator	Data source
Hazard	Rot index (potential for wood decay)	Met.No/Norwegian climate service Centre
	Flood hazard	Norwegian water resources and energy directorate
	Storm surge levels	Norwegian mapping authority
	Days with snow cover >30 cm	Norwegian climate service Centre
Vulnerability	Population density	Statistics Norway
	Vulnerable populations	Statistics Norway
	Business and industry exposure	Statistics Norway
	Mobility and infrastructure	NILU/environment agency/Institute of Transport Economics
Exposure	Storm—Natural hazard compensation	Finance Norway
	Flood—Natural hazard compensation	Finance Norway
	Storm surge—Natural hazard compensation	Finance Norway
	Landslide—Natural hazard compensation	Finance Norway
	Urban flooding/surface water compensation	Finance Norway
Response	Local adaptation measures with state funding	Norwegian environment agency
	Climate risk assessment in municipal planning	Norwegian directorate for civil protection
	Implementation of local adaptation measures funded locally	Norwegian climate monitor
	Integration of adaptation concerns with other policy areas	Norwegian climate monitor

- *Red*: Municipalities that have both made less effort to adapt and are more affected than other municipalities.

Our ranking yielded several key insights. Fortunately, one might say, it showed that the municipalities that are most heavily impacted are also those exhibiting the greatest response, as illustrated by the clustering in the upper-right quadrant of the cross in Fig. 6.7. There are practically no municipalities in the lower-right quadrant (high impact/low response).

The shifting vertical lines in the figure demonstrate how the overall climate impact increases through the 2000–2050 period and further to 2100. Geographically, coastal municipalities and those in Northern Norway consistently exhibit the highest risk. The ranking also shows that large urban municipalities, with their high density of

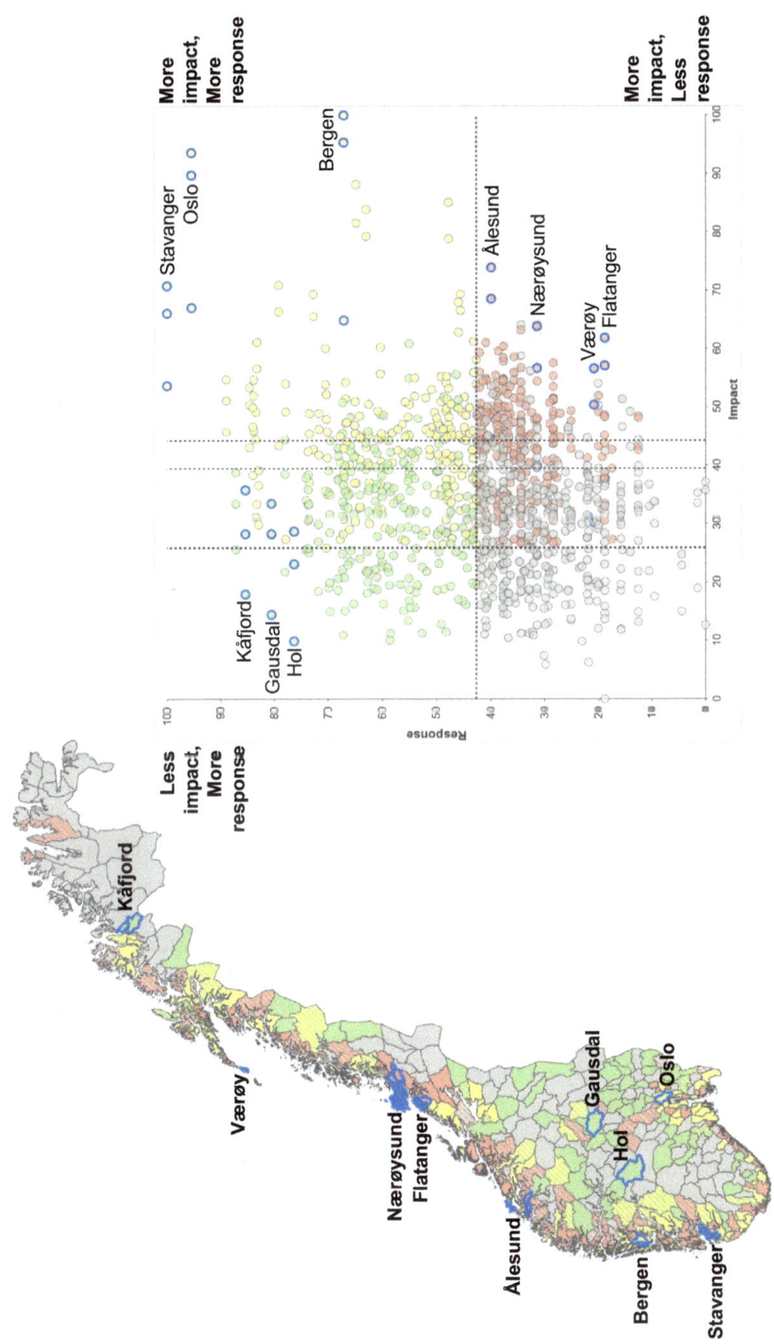

Fig. 6.7 Ranking of Norwegian municipalities in four levels of physical climate risk in 2023. Each point represents one municipality. The vertical white lines illustrate how the average impact increases from the year 2000 to 2050 and 2100

physical infrastructure and economic values (i.e., highest exposure), generally come out with higher overall risk than more rural municipalities, even though the latter group (especially coastal and mountain communities) tend to experience more severe climatic changes.

It is important to note that these insights were not in any way the result of predefined or ready-programmed conclusions. Rather, they were genuine findings that emerged only once the programming was finalized and the system was allowed to process the data independently. The fact that these autonomous results aligned so well with our expert understanding of the field provided significant reassurance that we were on solid scientific ground—even while acknowledging the methodological challenge of indexing such diverse sets of data. Moreover, the ranking system has been well received, and the fact that county authorities are commissioning in-depth analyses shows that Noradapt has successfully filled the gap where the state's official indicator projects had reached their limit (Nordland county[61] and Troms county[62]).

Beyond Borders: National Security and Transboundary Climate Risks

When I began researching the adaptation aspect of climate policy in the mid-2000s, I brought perspectives from my previous work on calculating GHG emissions. In the early 2000s, I participated in developing the first inventory of indirect emissions triggered by Norwegian consumption, demonstrating that consumption-based emissions were both greater and more negative in their development than territorial ones. In 2007, I presented the idea of transferring this dichotomy—within versus beyond borders—to climate change adaptation. This distinguished between direct and locally induced risk on the one hand, and indirect vs. internationally induced risk on the other.

To highlight the characteristics of this new form of risk, I created a model (or more correctly, a visualization) using the dependence of Norwegian livestock production on soybean imports from Brazil. The model highlights two critical points: first, that risk affects both production nodes and transport links; and second, that this version of transboundary physical climate risk is decoupled from local climate hazards but strongly influenced by climate hazards as well as exposures and vulnerabilities abroad.

The question of how climate change elsewhere affects Norway, which was raised at the start of our national debate, has been characterized by a divide between intellectual recognition and limited political acceptance. Since our 2009 analysis for Fredrikstad first identified food security and international conflict as key transboundary

[61] Klimamonitor (2026a).

[62] Klimamonitor (2026b).

risks,[63] the scientific basis has only grown stronger. Yet, despite the 2010 government committee's skepticism regarding "expensive food or refugees," the subsequent decade has proved these concerns to be increasingly valid.[213]

This skepticism culminated in the 2013 *White Paper on Climate Change Adaptation*, with the government firmly rejecting the issue[64]:

> No assessment has been made of how Norway may be indirectly affected by the impacts of climate change in other countries and regions.

In the following years, however, the tone shifted. By 2018, a public inquiry concluded that indirect effects likely represent the *greatest* risk for Norway.[216] That year, a report by the consultancy firm for the Environment Agency, EY, warned of significant consequences for security and food policy.[65] This was followed by a 2022 study stating that in extreme cases, climate change, combined with other events, could necessitate food rationing even in Norway.[66] This was a dramatic reversal from 2010, when officials optimistically predicted that climate change would merely provide a basis for *increased* food production. By 2024, the Office of the Auditor General assigned the severe rating of "inadequate" to Norway's contingency planning for food security.[67]

Taking the Risk: The Birth of Transadapt

Despite growing international recognition of transboundary climate risks, getting the Norwegian funding system to prioritize them was an uphill battle. We submitted a proposal entitled "Making Transborder Climate Risks Tangible for Adaptation Actions" (Transadapt) for what would have been the first major Norwegian research project on transboundary climate risk. The evaluation from the Research Council of Norway was excellent—a score of 6 out of 7. Yet, it was not funded.

Frustrated by this gap between scientific recognition and financial support, Noradapt made a bold strategic choice. We decided to use a significant portion of our own earmarked funding to finance a scaled-down version of the project ourselves. This became Transadapt. By taking this financial risk on our own account, we managed to make transboundary climate risks tangible for Norwegian decision-makers, with the Norwegian Environment Agency participating in the project's reference group.

Through several surveys conducted as part of the Norwegian Climate Monitor (Norsk klimamonitor) service operated by Noradapt,[68] we have demonstrated that

---

[63] Sælensminde et al. (2010).

[64] Ministry of Climate and Environment (2013).

[65] Prytz et al. (2019).

[66] Bardalen et al. (2022).

[67] Riksrevisjonen (2023).

[68] Klimamonitor (2026d).

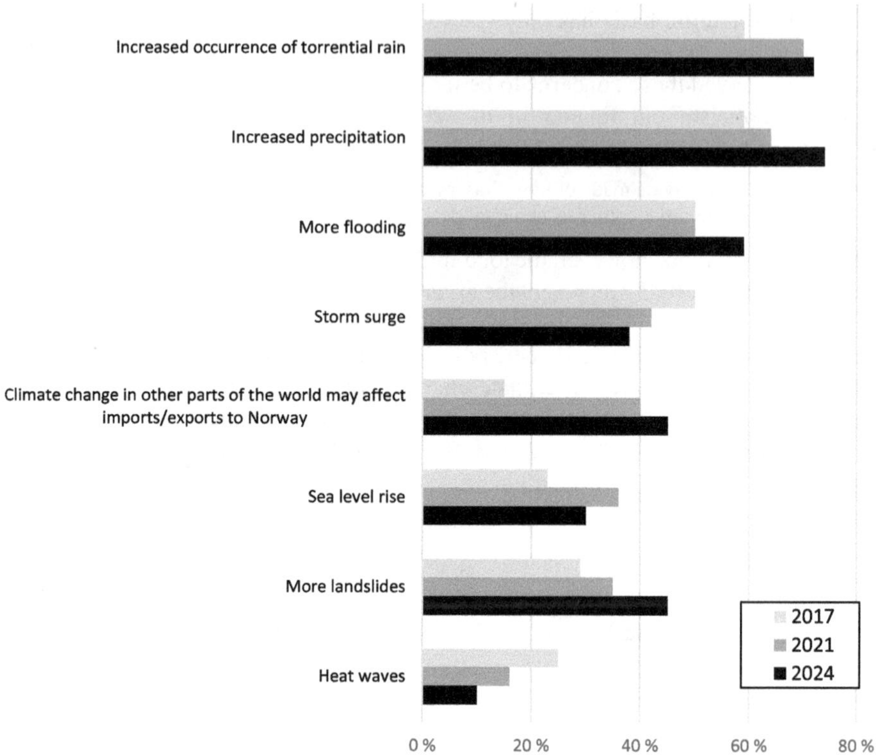

Fig. 6.8 Proportion of Norwegian municipalities that report being "very concerned" or "concerned" about eight types of climate change for the years 2017,[69] 2021, [265] and 2024[70]

concern about transboundary climate risk has increased more than any other risk category among those working with climate change adaptation. As shown in Fig. 6.8, the longitudinal data from Norwegian municipalities reveal a dramatic shift in perception. In 2017, only 15% of municipalities were concerned to a large or very large degree about how climate change in other parts of the world could affect imports and exports to Norway. By 2024, this figure had tripled to 45%. This shift was undoubtedly accelerated by the COVID-19 pandemic, which served as a brutal proof of concept for systemic vulnerability. Thus, the rapid increase in awareness has placed transboundary climate risk on par with major local physical hazards, such as landslides, in the minds of local decision-makers.

The second Norwegian parliamentary report on climate change adaptation—presented ten years after the first—partially reflects the signals from the numerous government studies published in the intervening period. It now acknowledges that transboundary climate risk is a relevant issue, with reference to food security. The

---

[69] Norwegian Association of Local and Regional Authorities (KS) (2018).

[70] Tandberg and Selseng (2024).

following year, in connection with the consideration of a parliamentary report on agricultural policy, the Storting adopted a goal to increase Norwegian self-sufficiency from 35% to 50%.[71] In 2025, the same target was included as one of four main objectives for strengthening security in a parliamentary report on preparedness.[72]

Nevertheless, this aspect of the climate issue remains a low-priority area within climate change adaptation policy, which itself is the least prioritized of the two branches of climate policy. This persists despite a growing number of studies indicating that transboundary climate risk poses a greater economic challenge for Norway—and the other Nordic countries—than local physical climate risk.[73]

Integrating Research on the Climate and Nature Crises

In 2010, my late colleague Karl Georg Høyer published an article that made an important contribution to moving climate policy out of its dead end and onto a new, more transformative path. The article was titled "Seven Theses on CO_2-Reductionism and Its Interdisciplinary Counteraction". In the article, he presents a hypothesis on climate reductionism that still resonates today (p. 35)[74]:

> The point of departure is the claim that the contemporary discourse on climate change is dominated by CO_2 reductionism. This form of reductionism reduces complex phenomena involving both nature and society to a single issue: CO_2 emissions.

Karl Georg identified three forms of reductionism: GHGs are reduced to CO_2, the energy issue is reduced to CO_2, and environmental issues are reduced to CO_2.

In the spring of 2020, my colleague and friend Ole Jørgen Grann from the KS contacted me. He shared a frustration that echoed the mechanisms Karl Georg had described ten years earlier: environmental considerations were continually being overshadowed by climate concerns. This initiative formed the basis for a project with the self-explanatory title "Knowledge Summary for KS on Interaction Effects in Local Climate and Environmental Policy". The project resulted in a model illustrating the numerous interconnections between four policy areas: climate change mitigation, energy transition, climate change adaptation, and biodiversity conservation. In line with our tradition of adopting a critical research perspective, we focused on the conflict dimension—hence the use of the minus sign ("÷") in Fig. 6.9. We concluded that biodiversity was most often the loser, while energy expansion emerged as the winner.

Our report to the KS was well received and has since been widely used in their professional and political work. The work we conducted for the KS also provided a scientific foundation for several research projects concerning the interface between

[71] Landbruks- og matdepartementet (2024).

[72] Justis- og beredskapsdepartementet (2025).

[73] Berninger et al. (2022).

[74] Høyer (2010).

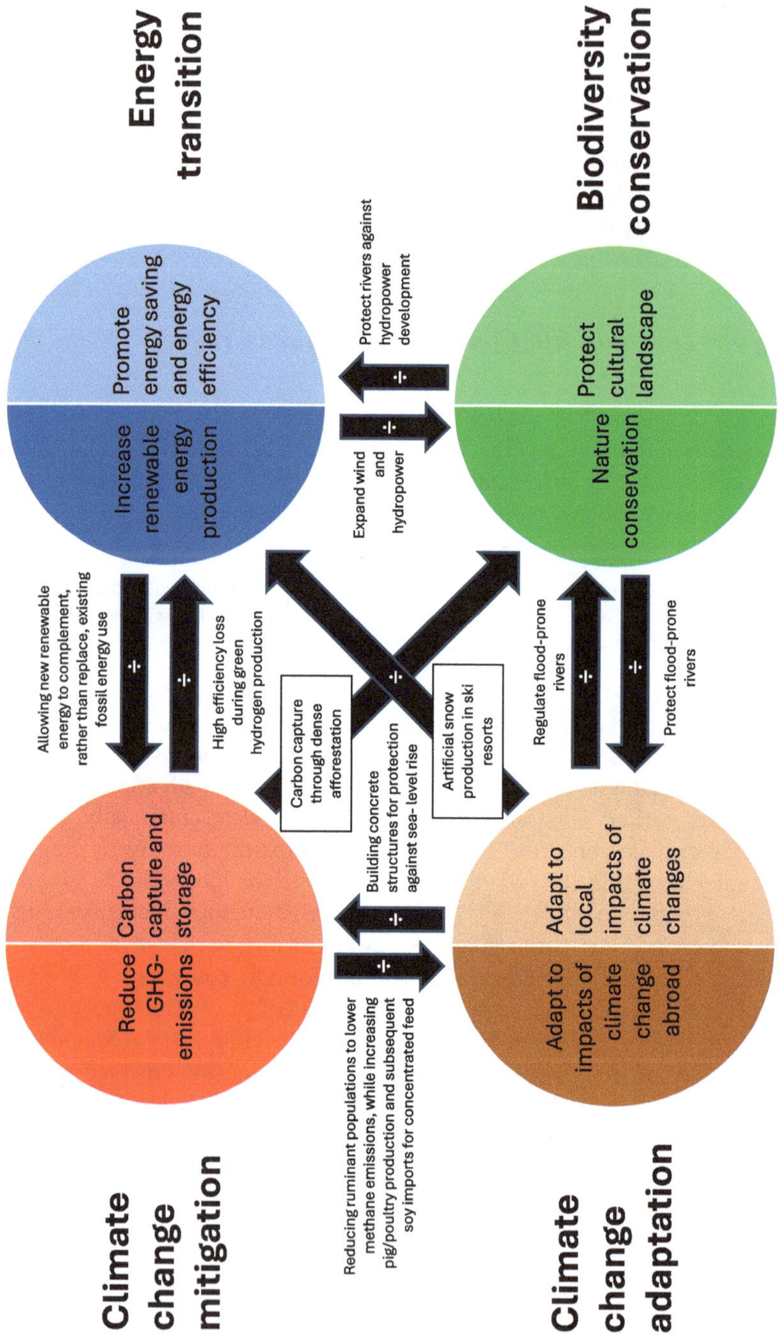

Fig. 6.9 Possible negative interaction effects between four related policy areas[75]

[75] Brendehaug et al. (2022).

climate considerations and biodiversity conservation. This new research direction culminated in the autumn of 2025 when WNRI—with my highly skilled colleague Irmelin Gram-Hanssen as director and myself as deputy director—submitted an application to the Research Council of Norway together with six Norwegian and one international research institute for an eight-year research center called the Norwegian Centre for Sustainable Land Use Through Transformative Research and Practice (Landfutures). In the final stages of preparing this book, we received the exhilarating news that the grant had been awarded.

The Fight Against Biodiversity Loss Is Rising on the Local Political Agenda

Since 2016, WNRI, in collaboration with several regional partners, has organized an annual national conference in Sogndal entitled "Climate Transformation," which each year gathers 100 to 150 participants from public administration and research communities to listen to and discuss the latest developments in research and best practices related to mitigating and adapting to climate change, both framed within the context of sustainable development. In recent years, we have included topics related to the protection of biodiversity (see www.klimaomstilling.no).

At the 2024 conference—a few months after the first reports from NRK's climate editorial team on biodiversity loss—I moderated a panel discussion between Håvard Nyhus (one of the journalists from the team), Truls Gulowsen (head of the Norwegian Society for the Conservation of Nature), and Marit Kasin (journalist at *Avisa Valdres* and author of the bestselling book *The Nature Paradox: About the Nature We Are Losing and How We Can Get It Back*). All three agreed that the numerous NRK reports on biodiversity loss have clearly influenced the environmental policy debate in Norway. Truls Gulowsen also noted that they had received reports from their local chapters across the country indicating that municipalities are showing a marked increase in interest in prioritizing the fight against biodiversity loss on the local political agenda.

Recent empirical data strongly supports Gulowsen's observation, demonstrating that this perceived shift in local interest is indeed translating into concrete municipal action. An analysis of public procurement data from the Norwegian national notification database for public procurement (Doffin) between 2024 and 2026 reveals a distinct transition in local governance. In 2024, municipal tenders related to environmental and planning sectors were still predominantly climate- and energy-centric, with biodiversity and land-use accounting for approximately 41% of the announcements ($n = 360$). By 2025, this relationship inverted ($n = 380$), and by the first half of 2026, tenders focusing on biodiversity conservation, land-use neutrality, and ecological mapping grew to constitute 60% of the total environmental procurement volume ($n = 120$).[76]

The Status of Climate Change Adaptation Efforts in Norway

The Noradapt leadership group has closely monitored the process surrounding Norway's most recent *White Paper on Climate Change Adaptation.* Throughout this process, we have emphasized five urgent challenges that have intensified since the previous report in 2013. These were presented in an op-ed published shortly after the white paper was released, but a few months before its deliberation in the Storting[77]:

- *Low priority*: Adaptation remains a very low priority in Norwegian environmental policy.
- *Infrastructure backlog*: There is a large and increasing backlog in maintaining and safeguarding physical infrastructure.
- *Transboundary blind spots*: National targets still ignore impacts on Norway of climate change occurring outside Norway, despite this being the risk with the greatest potential economic impact.
- *Lack of integration*: Adaptation must be integrated with emission reductions, renewable energy expansion, and biodiversity conservation.
- *Transformation gap*: Norway is failing to shift from incremental efforts to the radical transformations recommended by the IPCC.

In Fig. 6.10, I have tried to illustrate the mechanism of "decoupling emissions, attaching risk" by compiling data on CO_2 emissions, insurance loss-data for climate-related damage, and GDP. The results reveal a troubling trend: while Norway has to some extent decoupled its economy from emissions—generating more value with lower CO_2 output—we have simultaneously seen an "attachment" to climate risk. For every million kroner of value produced today, we face increasingly higher costs from extreme weather events and water intrusion.

International literature often characterizes climate change adaptation as "the poor cousin" in the climate family.[78] This is certainly the case in Norway. Figure 6.11

[76] Authors' own calculations based on data extracted from the Doffin database (The Norwegian National Notification Database for Public Procurement), managed by the Norwegian Agency for Public and Financial Management (DFØ). Data for the period January 2024 to May 2026 was extracted using the filters for CPV codes 90,700,000 (Environmental services) and 71,400,000 (Urban planning and landscape architectural services) for regional and local authorities. Available at: https://www.doffin.no/data

[77] Aall et al. (2023b, June 21).

[78] Palutikof et al. (2013).

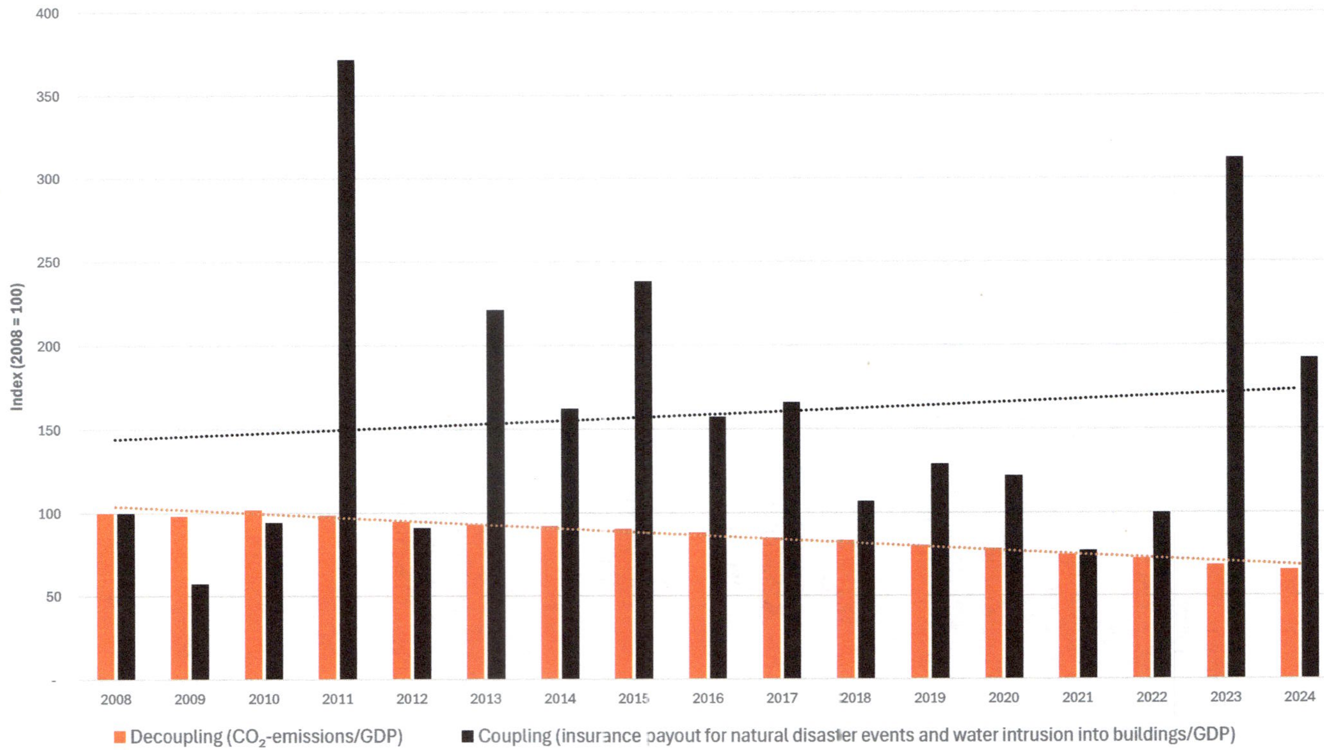

Fig. 6.10 Norwegian figures for emission intensity (CO_2/GDP) and damage intensity (natural damage + water intrusion)/GDP measured in 2023 NOK and indexed to 2008 = 100. Emission data[79] and gross domestic product (GDP)[80] are sourced from Statistics Norway, and insurance claims data are sourced from Finance Norway[81]

[79] Statistics Norway (2026a).

[80] Statistics Norway (2026b).

[81] Finance Norway (2026).

shows that despite a slight increase in funding, adaptation research accounted for a modest 8.7% of climate research funding in 2023. The clear winner remains renewable energy—a curious priority given that Norway's share of renewable primary energy use has remained stagnant at around 72% since 1990. Even more strikingly, this share had not increased leading up to 2024, whereas the global average had doubled during the same period.[82]

There is no comprehensive overview of the total resource allocation in Norway regarding policy measures and instruments in the climate sector. An indication of how national authorities prioritize local climate efforts can be seen in the Norwegian Environment Agency's scheme since 2016 for supporting local climate initiatives. These figures suggest an even greater imbalance in the allocation of funds between adaptation and emission reductions than in climate research, with only 3% of the total resources devoted to climate change adaptation over the entire period (see Fig. 6.12).

Ever since the first official report on climate change adaptation was published in 2010, the importance of addressing the maintenance backlog in physical infrastructure has been emphasized. In Fig. 6.13, I have compiled figures from a wide range of sources employing different methods for calculating the maintenance backlog, which may also vary from year to year. The figure nevertheless provides a basis for identifying key trends in this development: the maintenance backlog has increased, not decreased, since the issue was first linked to climate change in 2010, and growth has been substantial, with a total increase of nearly 400% over 15 years for all categories shown in the figure.

As it stands, Norway occupies a unique international position regarding the financial resources available to address the challenges posed by climate change. The country largely has the financial capacity to solve the immediate challenges that climate change creates for physical infrastructure by following two main strategic paths:

- *Accepting the risk and paying for damages*: Relying on robust compensation schemes to cover the costs when natural hazard events occur.
- *Buying our way out of the risk*: Investing heavily to eliminate the massive maintenance and safeguarding backlog of physical infrastructure.

Norway possesses one of the world's most robust compensation schemes for natural hazard events. It consists of a private component—financed through a mandated share of all fire insurance policies—and a public component, the Norwegian Natural Perils Compensation Scheme, which covers objects that cannot be insured privately. During the period shown in Fig. 6.13, insurance companies alone paid out 35 billion NOK in compensation for damage caused by floods, storms, landslides, and water intrusion.

However, the cost of *preventing* such damage is far higher. In 2021, the Norwegian Water Resources and Energy Directorate (NVE) estimated that securing existing private residences against floods and landslides would cost between 50 and 120 billion NOK. Crucially, this is a minimum estimate: it excludes holiday homes, commercial

[82] Our World in Data (2026b).

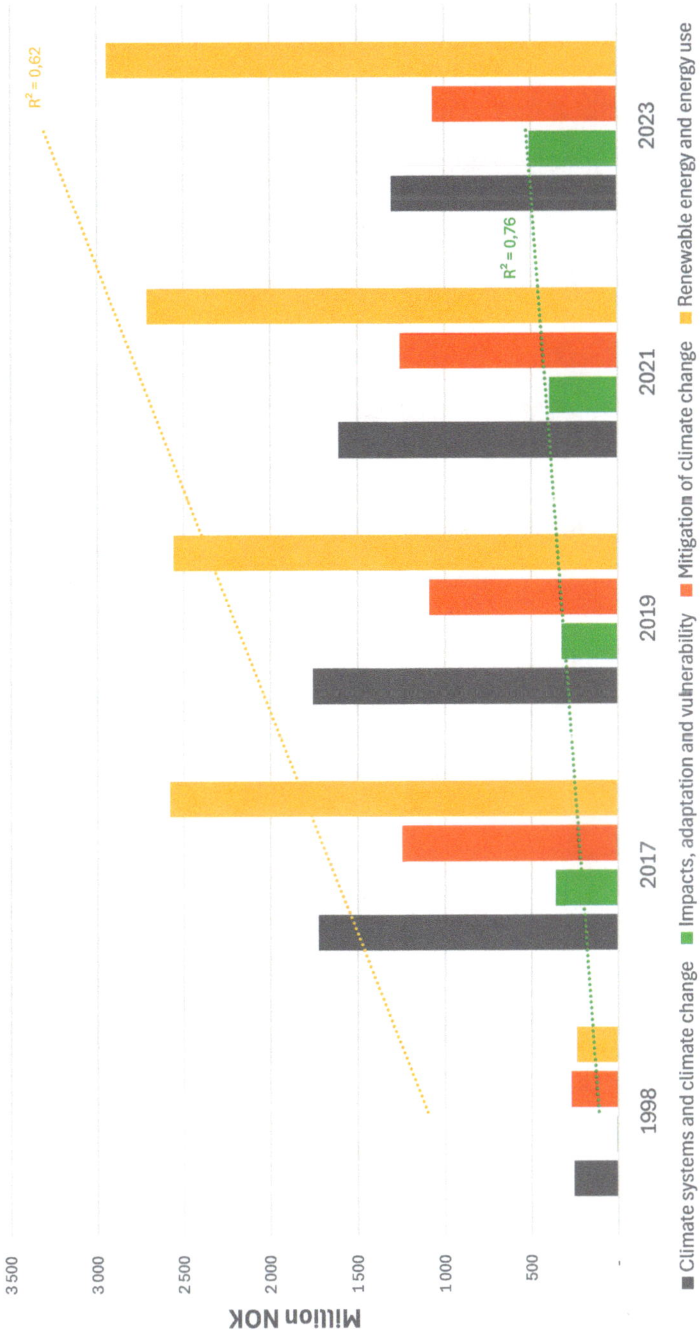

Fig. 6.11 Distribution of allocations from the Research Council of Norway for climate research from 1998 to 2023 measured in constant 2023 prices[83]

[83] Research Council of Norway (2026).

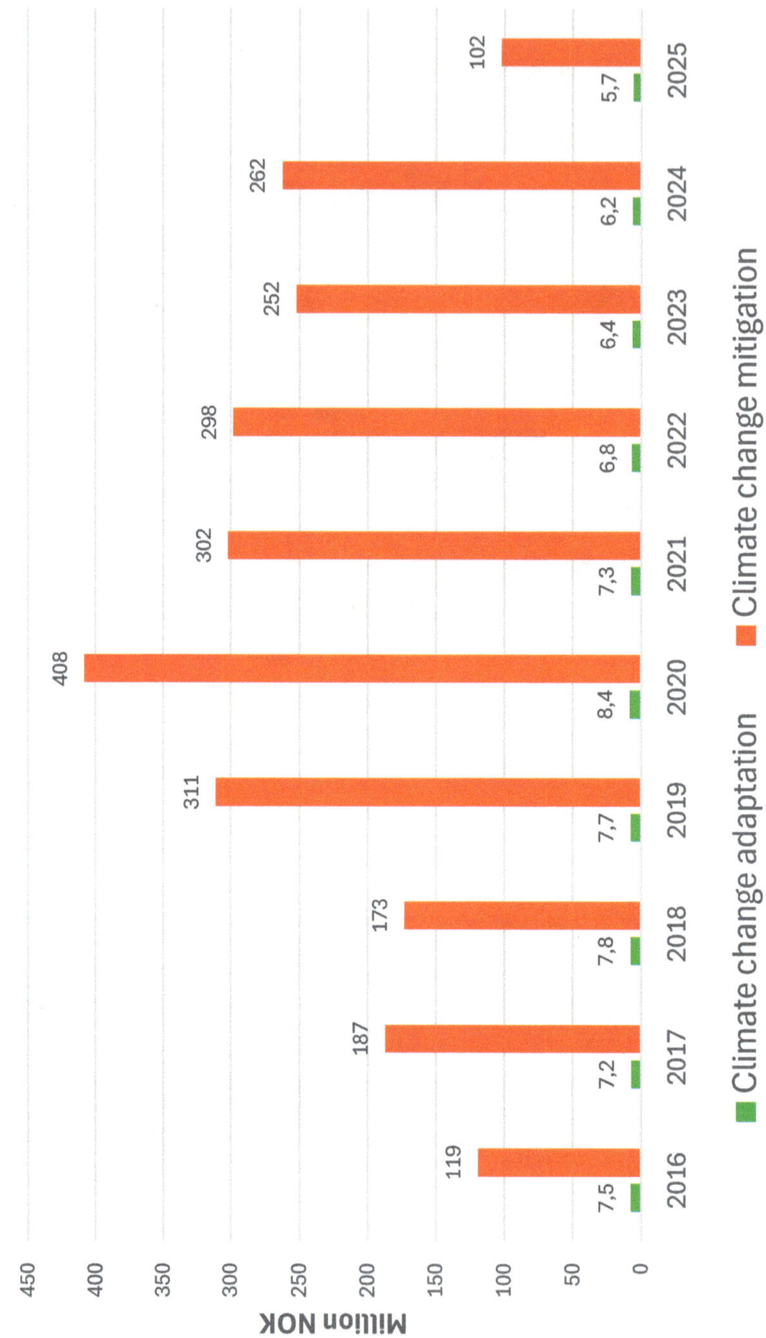

Fig. 6.12 Allocation of grants for local climate initiatives from the Norwegian Environment Agency related to climate change adaptation and emission reductions from 2016 to 2025, measured in constant 2023 prices

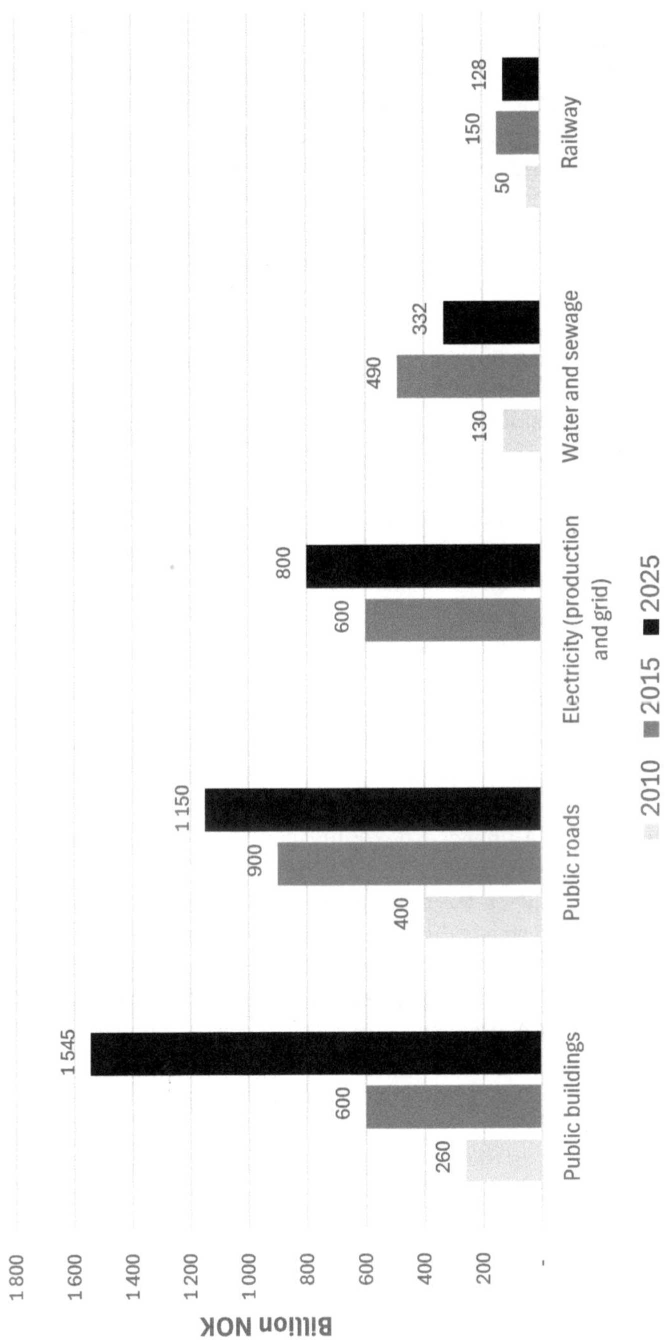

Fig. 6.13 Maintenance backlog for various categories of physical infrastructure in Norway for 2010, 2015, and 2025 measured in constant 2023 prices, compiled from data from NOU 2010,[210] National Transport Plans,[84] the Norwegian Association of Consulting Engineers,[85] and the municipal–state reporting system (KOSTRA)[86]

84 Norwegian Public Roads Administration (2026).
85 Rådgivende Ingeniørers Forening (RIF) (2025).
86 Statistics Norway (2026c).

buildings, and all public infrastructure like roads and railways. It also assumes that all buildings constructed after 2021 are safe and, perhaps most importantly, it does not account for the intensified risks brought by future climate change.

The scale of these costs must be viewed in the context of Norway's unique financial position. The extreme flood "Hans" in 2023 resulted in insurance payouts of at least 8.5 billion NOK.[87] While a significant sum, it corresponds to just under three days (2 days and 21 h, to be precise) of the Norwegian state's cash flow from oil and gas production taxes that same year.[88] It is reasonable to assume that decision-makers have operated under the assumption that Norway can simply afford to pay for damages as they occur, rather than bearing the political and financial effort of large-scale prevention.

If the Norwegian state decided to "buy its way out" of climate risk, how far would our accumulated wealth go? To place the "buy-out" option in a broader perspective, we must look at the state's overall financial capacity. While the need for securing residential buildings is estimated at up to 120 billion NOK, the total maintenance backlog for physical infrastructure is approximately 3300 billion NOK (as shown in Fig. 6.13).

At the beginning of 2025, the Government Pension Fund Global (GPFG) reached a value of approximately 19,000 billion NOK. Since 2001, Norway has followed the "fiscal rule", a commitment to spend no more than the annual expected return from the fund (currently set at 3%). If we assume this rule is upheld, addressing both the infrastructure maintenance backlog and the safety requirements for private homes would require an additional withdrawal from the fund equivalent to only 0.25% per year over a 15-year period.

The question, therefore, is not whether the Norwegian economy has the capacity to manage the portion of climate risk tied to the exposure of physical infrastructure. Even with a gradual decline and eventual halt in oil production toward the year 2100, our accumulated wealth provides a financial flexibility that surpasses almost any other nation.

However, climate risk encompasses far more than what is created by "local" climate changes impacting "local" physical infrastructure. Economists have calculated that in the long term, transboundary climate risk likely represents the greatest economic threat to the Norwegian economy. Furthermore, these calculations do not yet account for the more creeping consequences of climate change on nature and ecosystems—simply because we do not know how these impacts will manifest at a systemic level. Consequently, we cannot yet quantify the economic fallout of, for example, a collapse in economically vital fish stocks along the Norwegian coast due to a warmer and more acidic ocean, or the dieback of commercially managed spruce forests in parts of Norway that will eventually become too warm and dry for the species.

The real question is whether political priorities will shift from "consumption today" toward the long-term safeguarding of our physical infrastructure and natural

[87] Aall and Jansen (2024).

[88] Norsk Petroleum (2026).

foundations. As it stands, the risks associated with a changing climate extend far beyond our current estimates. We still lack credible assessments of the costs related to diffuse climate impacts on ecosystems, or the economic weight of transboundary risks—both of which are likely to dwarf the direct costs of local physical infrastructure damage.

References

Aaheim, A., Dannevig, H., Ericsson, T., van Oort, B., Innbjør, L., Rauken, T., Vennemo, H., Johansen, H., Tofteng, M., Aall, C., Groven, K., & Heiberg, E. (2009). *Konsekvenser av klimaendringer, tilpasning og sårbarhet i Norge* [Consequences of climate change, adaptation and vulnerability in Norway] (CICERO Report 2009:04). CICERO.

Aall, C. (2005a). *Personal notes from the conference "Recruitment of Social Sciences into Climate Research."* Research Council of Norway.

Aall, C. (2005b). *Personal notes from the regional climate change adaptation project in Sogn og Fjordane* (2004–2005).

Aall, C. (2011a). *Personal notes from seminar with the Bergenshus Home Guard* (HV-09). Bergen.

Aall, C. (2023a, August 7). Klimaskeptikerne har litt rett. *NRK Ytring*. https://www.nrk.no/ytring/klimaskeptikerne-har-litt-rett-1.16502454

Aall, C. (Ed.). (2011b). *Klimaendringenes konsekvenser for kommunal og fylkeskommunal infrastruktur. Sluttrapport* [Consequences of climate change for municipal infrastructure. Final report] (VF-report 4/2011). Western Norway Research Institute.

Aall, C. (Ed.). (2023b). *Protocol for the case studies of the UNCHAIN project*. (VF-note 2/2023). Western Norway Research Institute.

Aall, C., & Groven, K. (2003). *Institusjonell respons på klimaendringer* [Institutional response to climate change] (VF report 3/03). Western Norway Research Institute.

Aall, C., & Groven, K. (2022). The unpredictable truth: A proposed road map for a reflect-then-act approach to climate uncertainties and lessons learned from Norwegian municipalities. *Weather, Climate, and Society, 14*(1), 337–347.

Aall, C., & Jansen, M. K. (2024). Skuldast skadane etter ekstremvêret «Hans» dårleg vêr eller dårleg planlegging? [Is the damage caused by extreme weather event "Hans" due to bad weather or poor planning?] In D. Grønnestad (Ed.), *Hvem Hva Hvor 2025*. Vigmostad & Bjørke.

Aall, C., & Norland, I. T. (2003). *Indikatorer for vurdering av lokal klimasårbarhet* [Indicators for assessing local climate vulnerability] (VF-report 15/03). Western Norway Research Institute.

Aall, C., Aalen, P., Bruvoll, A., Eidsvig, U. (2025a). *Sluttrapport fra innovasjonsprosjektet «klimatilpassing og veitransport»*. [Final report from the innovation project "Climate Adaptation and Road Transport"] (VF-report 3/2024). Western Norway Research Institute.

Aall, C., Botnen Holm, T., Cauchy, A., Rudolf, F., Harris, K., Jansen, M. K., Gobert, J., Lager, F., Arvis, B., & Bour, M. (2023a). Think global—act local: The challenge of producing actionable knowledge on transboundary climate risks for the local level of governance. *Frontiers in Climate, 5*. https://doi.org/10.3389/fclim.2023.1170142

Aall, C., Chang, M., Holm, T.B., Løkke, S., Mati, A., Mayer, S., & Skov, R. (2025b). Overlooked climate risks in the ongoing renewable energy transitions. *Energy, 335*, Article 137986. https://doi.org/10.1016/j.energy.2025.137986

Aall, C., Dannevig, H., Brendehaug, E., Kvitsand, H., Rød, J. K., Vindegg, M., Neby, S., Pontoppidan, M., Hovelsrud, G., & Dahle, L. (2023b, June 21). Er Norge forberedt på tørkesommeren? *Nationen*. https://www.nationen.no/er-norge-forberedt-pa-torkesommeren/o/5-148-378285

Aall, C., Gerger Swartling, Å., & Attoh, E. M. (2023c). Editorial: New approaches to local climate change risk analysis. *Frontiers in Climate, 5*, Article 1298779. https://doi.org/10.3389/fclim.2023.1298779

Aall, C., Wanvik, T., & Dale, B. (2022). Climate risks of the transition to a renewable energy society: The need for extending the research agenda. *Weather, Climate, and Society, 14*(2), 387–397. https://doi.org/10.1175/WCAS-D-21-0055.1

Bardalen, A., Pettersen, I., Dombu, S. V., Rosnes, O., Mittenzwei, K., & Skulstad, A. (2022). *Klimaendring utfordrer det norske matsystemet*. NIBIO/Norwegian Environment Agency.

Berninger, K., Lager, F., Botnen Holm, T., Tynkkynen, O., Klein, R. J. T., Aall, C., Dristig, A., Määttä, H., & Perrels, A. (2022). *Nordic perspectives on transboundary climate risk: Current knowledge and pathways for action* (TemaNord 2022:531). Nordic Council of Ministers.

Breivik, R., Aall, C., & Rød, J. K. (2014). *Pilotprosjekt om testing av skadedata fra forsikringsbransjen for vurdering av klimasårbarhet* [Pilot project on testing damage data from the insurance industry for climate vulnerability assessment] (VF-report 7/2014). Western Norway Research Institute.

Brendehaug, E., Groven, K., Selseng, T., & Aall, C. (2022). *Samspeleffektar i lokal miljø- og klimapolitikk* [Synergy effects in local environmental and climate policy] (VF-report 4/2022). Western Norway Research Institute.

Buan, I. F., Heggelund, G., Jevnaker, T., & Rosendal, K. G. (2010). *Norge og internasjonale ringvirkninger av klimaendringer: En forstudie* (FNI Report 11/2010). Fridtjof Nansen Institute.

Dannevig, H., Beitnes, S., & Tanberg, B. (2024). *Forslag til metode for nasjonal klimasårbarhetsanalyse* [Proposal for a method for national climate vulnerability analysis]. (VF-report 12/2024). Western Norway Research Institute.

Dannevig, H., Groven, K., & Aall, C. (2015). *Er dagens kommunale arealplanlegging i stand til å sikre samfunnet mot klimaendringar?* [Is current municipal spatial planning capable of securing society against climate change?] (VF-report 8/2013). Western Norway Research Institute.

Dannevig, H., Groven, K., & Aall, C. (2016). *Oktoberflaumen på Vestlandet i 2014* [The October flood in Western Norway 2014] (VF-report 3/2016). Western Norway Research Institute.

Dannevig, H., Groven, K., Aall, C., & Brevik, R. (2013). *Kva kan vi lære av historiske naturskadehendingar for betre tilpassing til klimaendringar?* [What can we learn from historical natural hazard events to better adapt to climate change?] (VF-report 10/2014). Western Norway Research Institute.

Dannevig, H., Heiberg, E., Aall, C. (2014). *Analyse av klimasårbarhet Arbeidsbok til bruk i det lokale klimatilpasningsarbeidet i Sogn og Fjordane* [Climate vulnerability analysis workbook for use in local climate adaptation work in Sogn og Fjordane] (VF-note 7/2014). Western Norway Research Institute.

DSB. (2026). *Kunnskapsbanken* (The Knowledge Bank). https://www.dsb.no/statistikk/natur%2D%2Dog-vannskader/

Finanace Norway. (2026). https://www.finansnorge.no/tema/statistikk-og-analyse/forsikring/natur%2D%2Dog-varskader/

Fritzsche, K., Schneiderbauer, S., Bubeck, P., Kienberger, S., Buth, M., Zebisch, M., & Kahlenborn, W. (2014). *The vulnerability sourcebook. Concept and guidelines for standardized vulnerability assessments.* GIZ

Groven, K. (2005). *Klimasårbarhet i bustadsektoren—Lokal sårbarheitskartlegging og klimatilpassing*. [Climate vulnerability in the housing sector—Local vulnerability mapping and climate change adaptation] (VF-rapport 1/2005). Western Norway Research Institute.

Groven, K., Aall, C., van den Berg, M., Kanyama, A. C., & Coenen, F. (2012). Civil protection and climate change adaptation. Comparing lessons learned from three coastal cities in Norway, Sweden and The Netherlands. *Local Environment,* 17(6–7), 679–694.

Groven, K., Leivestad, H. H., Aall, C., Selstad, T., Høydal, Ø. A., Nilsen, A. S., & Serigstad, S. (2008). *Naturskade i kommunene—Sluttrapport fra prosjekt for KS* [Natural damage in municipalities—Final report from the project for the KS] (VF-report 4/2008). Western Norway Research Institute.

Groven, K., Lundli, H.-E., & Aall, C. (1999). *Lokal klimapolitikk - internasjonale og nasjonale erfaringer* [Local climate policy - international and national experiences] (VF-report 4/99). Western Norway Research Institute

Holm, M., & Aall, C. (2022). *Klimarisiko for vegtransportsektoren* [Climate risk for the road transport sector] (VF-report 3-2022). Western Norway Research Institute.

Høyer, K. G. (2010). Seven theses on CO2-reductionism and its interdisciplinary counteraction. In R. Bhaskar, C. Frank, K. G. Høyer, P. Næss & J. Parker (Eds.), *Interdisciplinarity and climate change. Transforming knowledge and practice for our global future* (pp. 35–53). Routledge

Husabø, I. (2009). *Klimatilpassing i Sogn og Fjordane anno 2008* [Climate change adaptation in Sogn og Fjordane 2008] (VF-notat 4/2009). Western Norway Research Institute.

Husabø, I. (2010). *Ekstremvêrhendingar. Erfaringsgrunnlag for klimatilpassing hos fylkesmannen* [Extreme weather events. Experiences with climate change adaptation] (VF-report 4/2010). Western Norway Research Institute.

Intergovernmental Panel on Climate Change (IPCC). (2007). *Climate change 2007: Impacts, adaptation and vulnerability: Contribution of Working Group II to the Fourth Assessment Report of the Intergovernmental Panel on Climate Change* (M. L. Parry, O. F. Canziani, J. P. Palutikof, P. J. van der Linden, & C. E. Hanson, Eds.). Cambridge University Press.

Intergovernmental Panel on Climate Change (IPCC). (2022). *Climate change 2022: Impacts, adaptation and vulnerability. Contribution of Working Group II to the Sixth Assessment Report of the Intergovernmental Panel on Climate Change* (H.-O. Pörtner, D. C. Roberts, M. Tignor, E. S. Poloczanska, K. Mintenbeck, A. Alegría, M. Craig, S. Langsdorf, S. Löschke, V. Möller, A. Okem, & B. Rama, Eds.). Cambridge University Press.

Justis- og beredskapsdepartementet. (2025). Totalberedskapsmeldingen. Forberedt på kriser og krig (Meld. St. 9 [2024–2025]) [White paper].

Klimamonitor. (2026a). *Climate risk in Nordland county* [Story map]. https://klimamonitor.no/klimarisiko/nordland

Klimamonitor. (2026b). *Climate risk in Troms county* [Story map]. https://klimamonitor.no/klimarisiko/troms

Klimamonitor. (2026c). *The geography of future climate risk.* https://klimamonitor.no/

Klimamonitor. (2026d). www.klimamonitor.no

Landbruks- og matdepartementet. (2024). Strategi for auka sjølvforsyning av jordbruksvarer (Meld. St. 11 [2023–2024]) [White paper].

Levin, K., Cashore, B., Bernstein, S., & Auld, G. (2012). Overcoming the tragedy of super wicked problems: Constraining our future selves to ameliorate global climate change. *Policy Sci, 45*, 123–152. https://doi.org/10.1007/s11077-012-9151-0

Mehta, L., Leach, M., & Scoones, I. (2001). Editorial: Environmental governance in an uncertain world. *IDS Bulletin, 32*(4), 1–9. https://www.ids.ac.uk/files/dmfile/mehtaetal32.4.pdf

Miles, M., & Richter, K. (2011). *Klimaendringenes konsekvenser for kommunal og fylkeskommunal infrastruktur. Delrapport 2: Klimaanalyse.* [The consequences of climate change for municipal and county infrastructure. Sub-report 2. Climate analysis]. The Bjerknes Centre.

Miljødirektoratet. (2026). *Miljøindikator 5.5.1.* https://miljostatus.miljodirektoratet.no/miljomal/klima/miljomal-5.5/miljoindikator-5.5.1

Ministry of Climate and Environment. (2013). *A changing climate—United for a climate-resilient society* (Meld. St. 26 [2022–2023]) [White paper].

Moe, O. B. (2022, March 11). Statement during ministerial visit to Western Norway Research Institute. Personal communication.

Norges offentlige utredninger (NOU) (2010). *Tilpassing til eit klima i endring* [Adapting to a changing climate] (NOU 2010: 10). Departementenes servicesenter.

Norges offentlige utredninger (NOU). (2018). *Klimarisiko og norsk økonomi* [Climate risk and the Norwegian economy] (NOU 2018: 17). Departementenes servicesenter.

Norges vassdrags- og energidirektorat (NVE). (2019). *Utredning av sikkerhet mot skred i bratt terreng* [Assessing safety against landslides in steep terrain] (Guideline 1/2019).

Norsk Petroleum. (2026). *The state's revenues.* https://www.norskpetroleum.no/okonomi/statens-inntekter/

Norwegian Association of Local and Regional Authorities (KS). (2018). *Klimatilpasning i kommunene—nasjonal spørreundersøkelse for KS høsten 2017*. https://www.ks.no/contentassets/87146f43e9f346e18991ceb748236ac9/klimatilpasning_nasjonal_sporreundersokelse_april2018.pdf

Norwegian Centre for Climate Services (NCCS). (2011). *Climate profile for Sogn og Fjordane and Nordhordland: Basis for climate change adaptation.*

Norwegian Centre for Climate Services (NCCS). (2025). Climate projections for Norway: Temperature and precipitation scenarios. https://klimaservicesenter.no/climateprojections

Norwegian Public Roads Administration. (2026). https://www.vegvesen.no/nn/vegprosjekt/nasjonal-transportplan/tidligere-nasjonale-transportplaner/

Office of the Auditor General of Norway. (2022). *The Office of the Auditor General's investigation of the authorities' work on climate change adaptation of infrastructure and buildings* (Document 3:6 [2021–2022]). https://www.riksrevisjonen.no/globalassets/reports/en/2021-2022/climate-change-adaptation.pdf

Our World in Data. (2026a). *Energy production and consumption.* https://ourworldindata.org/energy

Our World in Data. (2026b). *Renewable energy statistics.* https://ourworldindata.org/renewable-energy

Palutikof, J. P., Boulter, S. L., Ash, A. J., Stafford Smith, M., Parry, M., Waschka, M., & Guitart, D. (2013). *Climate adaptation futures.* John Wiley & Sons.

Perrow, C. (2007). *The next catastrophe: Reducing our vulnerabilities to natural, industrial, and terrorist disasters.* Princeton University Press.

Prytz, E., Storhaug, F., & Fadnes, Y. (2019). *Utredning om kunnskap og håndtering av klimarisiko i andre land.* EY/Norwegian Environment Agency.

Rådgivende Ingeniørers Forening (RIF). (2025). *Norges tilstand 2025.* https://rif.no/norges-tilstand

Reisinger, A., Howden, M., Vera, C., Garschagen, M., Mach, K. J., Pathak, M., Poloczanska, E., van Aalst, M., Ruane, A. C., Hurlbert, M., Mintenbeck, K., Pedace, R., Rojas Corradi, M., Viner, D., Kreibiehl, S., O'Neill, B., Pörtner, H.-O., Sillmann, J., Jones, R., & Ranasinghe, R. (2020). *The concept of risk in the IPCC Sixth Assessment Report: A summary of cross-working group discussions.* Intergovernmental Panel on Climate Change.

Research Council of Norway. (2026). https://www.forskningsradet.no/siteassets/publikasjoner/1254013189546.pdf?utm_source=chatgpt.com, https://www.ssb.no/statbank/table/13711

Riksrevisjonen. (2021). *General's investigation into government authorities' effort to adapt infrastructure and built-up areas to a changing climate.* (Document 3:6 [2021–2022]). The Office of the Auditor General.

Riksrevisjonen. (2023). *Matsikkerhet og beredskap på landbruksområdet* [Food security and preparedness in the agricultural sector] (Document 3:6 [2021–2022]). Office of the Auditor General.

Rød, J. K., Aall, C., & Selseng, T. (2025). Towards a holistic climate service providing information about four climate risk determinants. *Climate Services, 38*, Article 100558. https://doi.org/10.1016/j.cliser.2025.100558

Sælensminde, I., Aall, C., & Hygen, H. O. (2010). *Klimatilpasning i Fredrikstad. Faglige innspill til Fredrikstad kommunes arbeid med en plan for tilpasning til Klimaendringer* [Climate change adaptation in Fredrikstad. Input to Fredrikstad Municipality's work on a plan for adaptation to climate change] (VF-report 3/2010). Western Norway Research Institute.

Sataøen, H. L., & Aall, C. (2006). *Introduksjon til klimasårbarhet og klimatilpasning i Fredrikstads klima- og energiplan.* [Introduction to climate vulnerability and climate change adaptation in Fredrikstad's climate and energy plan] (VF-note 16/2006). Western Norway Research Institute.

Selseng, T. (2026) (In review)

Selseng, T., Skogvang, B. J., & Aall, C. (2021). *Spørreundersøkelse til norske kommuner om status for 2021 i arbeidet med klimatilpasning* [Survey on climate adaptation status 2021] (VF-report 10/2021). Western Norway Research Institute.

Selstad, T. (2008). *Norge gjennom hundre år* [Norway through a hundred years] (ØF-rapport 03/2008). Østlandsforskning.

Simonsen, M., Kjønaas, O. J., & Aall, C. (2022). Substitution of fossil-energy intensive building materials by wood products—Does it matter? *Journal of Cleaner Production*, 383, Article 134941. https://doi.org/10.1016/j.jclepro.2022.134941

Statistics Norway. (2026a). https://www.ssb.no/statbank/table/09190

Statistics Norway. (2026b). https://www.ssb.no/statbank/table/13931

Statistics Norway. (2026c). KOSTRA. https://www.ssb.no/statbank/list/kostrahoved

Tandberg, B. K. R., & Selseng, T. (2024). *Status for kommunal klimatilpasning i 2024* [Status for municipal climate change adaptation 2024] (VF-report 4/2024). Western Norway Research Institute.

Teigland, J. (2002a). *Hvordan vær og klima påvirker reiselivs- og rekreasjonsatferd* [How weather and climate affect tourism and recreation behavior] (VF-report 11/2002). Western Norway Research Institute.

Teigland, J. (2002b). *Sosioøkonomiske effekter av ekstremt vær i Norge* [Socio-economic effects of extreme weather in Norway] (VF-report 7/2002). Western Norway Research Institute.

Verstat. (2026). https://verstat.no/

Wallgren, O. (2011). Nordic Adaptation 2010. In *Mistra-SWECIA annual report 2010* (pp. 16–17). Mistra-SWECIA, Swedish Meteorological and Hydrological Institute.

WWF Verdens naturfond. (2025). *Innspill til Ekspertutvalget for klimatilpasning* [Submission to the Expert Committee on Climate Change Adaptation]. Government.no. https://files.nettsteder.regjeringen.no/wpuploads01/sites/579/2025/12/WWF-Verdens-naturfond.pdf

Chapter 7
Conclusions and Recommendations on the Path Toward a Truly Sustainable Climate Transformation

Abstract This concluding chapter synthesizes 30 years of research on Norwegian environmental policy and local climate action. Beginning with the enduring echo of Franklin D. Roosevelt's "Look to Norway" speech—which historically cemented the perception of Norway as an exemplary nation—the chapter challenges the contemporary image of the country as an environmental pioneer by contrasting historical resource optimism with the current "piecemeal" degradation of nature. The author argues that the most powerful driver of transformation lies in Strategic Niche Management (SNM)—specifically through alliances between progressive institutional entrepreneurs within the state administration and pioneering municipalities. Using the Environmental Protection in Municipalities (*Miljøvern i kommunene* [MIK]) and Local Agenda 21 programs as primary examples, the chapter demonstrates how these protected spaces enabled policy innovations that top-down mandates could not achieve. Finally, the author highlights a unique strategic dimension of Local Agenda 21: its role as a democracy-builder from the bottom up, suggesting that such local, democratic free spaces can serve as a crucial counterbalance to national political inertia and global instability.

"Look to Norway?"

The slogan "Look to Norway" is frequently—and at times self-ironically—invoked by Norwegians whenever the country distinguishes itself on the international stage, whether it pertains to electric vehicle policy, peace mediation, the Winter Olympics, or our welfare schemes. The phrase originates from a speech delivered by U.S. President Franklin D. Roosevelt on September 16, 1942, during the handover of the submarine chaser *King Haakon VII* to the Royal Norwegian Navy.

The backdrop for the speech was the American press's portrayal of Norway following the 1940 German invasion, which had unfairly depicted Norwegians as offering little resistance, or as being passive or even treacherous. Roosevelt's speech was a deliberate effort to repair Norway's reputation, highlighting the nation as a

C. Aall, *Sustainable Development*, https://doi.org/10.1007/978-3-032-35930-8_7

symbol of heroic resistance while simultaneously justifying U.S. involvement in the war[1]:

> If there is anyone who still wonders why this war is being fought, let him look to Norway. If there is anyone who harbours any illusions that this war could have been avoided, let him look to Norway. And if there is anyone who doubts the democratic will to prevail, again I say, let him look to Norway.

The resonance of Roosevelt's words suggests a dual perspective: should one—or should one *not*—"look to Norway" when it comes to environmental policy? I share this sense of duality as I summarize and conclude my experiences from thirty years of research both "for" and "on" a sustainable climate transition. Should other countries "look to Norway" to learn how such a transition should—or perhaps should not—be organized?

Norway Misses Its Own Climate Targets Twice Over

Norway, alongside our Nordic neighbors Sweden and Denmark, and the environmental frontrunner the Netherlands, was among the first nations to adopt a national climate target. Norway has so far operated under six successive climate targets:

- *First target (1989)*: Stabilize CO_2 emissions at 1989 levels by the year 2000.
- *Second target (1997–2008)*: Maximum 1% increase in greenhouse gas (GHG) emissions from 1990 to 2010–2012.
- *Third target (2008–2012)*: 8% reduction of GHG emissions from 1990 to 2020 (with two-thirds achieved domestically).
- *Fourth target (2012–2018)*: 20% reduction of GHG emissions from 1990 to 2020 (with two-thirds achieved domestically).
- *Fifth target (2018)*: At least 40% reduction of GHG emissions by 2030 and 80–95% reduction by 2050 (from 1990 levels).
- *Sixth target (2021)*: 50–55% reduction of GHG emissions by 2030 and 90–95% reduction by 2050 (from 1990 levels).

Figure 7.1 illustrates a trend in which climate targets have become increasingly ambitious, while emission reductions have only weakly followed suit. The green trend line in the figure may appear to indicate that reductions are moving in the right direction, but a closer look at the data reveals that the trend has very low explanatory power—only 23%. This means that factors other than increasing climate ambitions influence actual emission cuts.

The significant percentage reduction in 1991 resulted as much from the banking crisis and economic recession as from Norway's introduction of a CO_2 tax on oil and gas production, which led to immediate emission reductions. The two major emission cuts in the early 2000s were primarily the result of fluctuations in the electricity

[1] Roosevelt (1942, September 16).

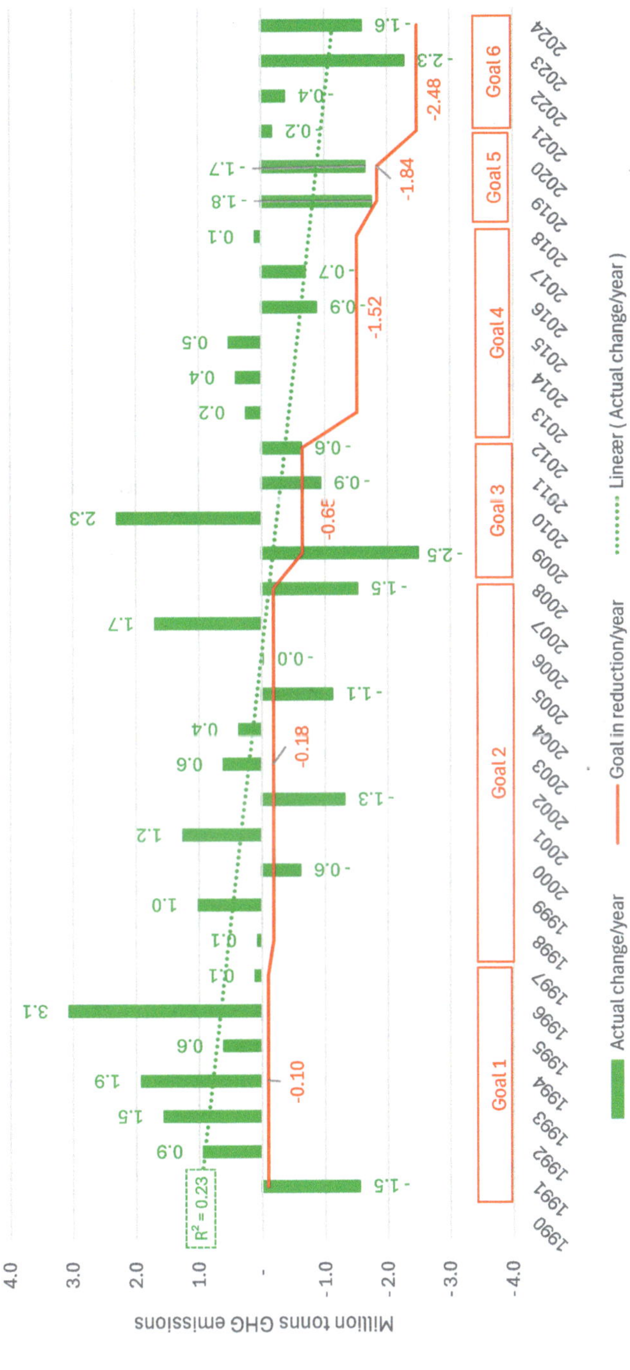

Fig. 7.1 Annual changes in Norwegian greenhouse gas (GHG) emissions compared with the prevailing national climate targets since 1990, where the total assumed emission reduction for each goal-period is distributed evenly per year

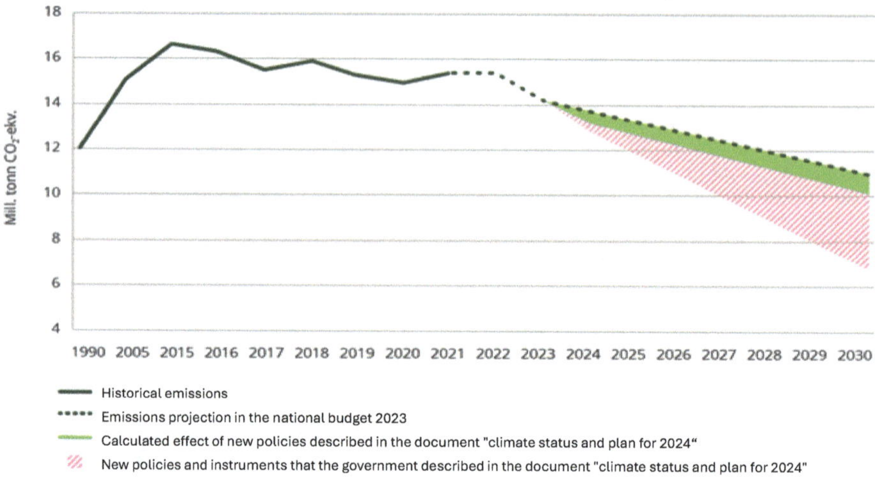

Fig. 7.2 Example of a "methodological error" frequently made by environmental authorities when presenting the desired development of greenhouse gas emissions from present to a future year (figure taken from a Norwegian government report)[2]

market caused by cold winters and low precipitation (which increased the need for fossil fuel-based peak load in heating and industry), as well as the closure of certain energy-intensive industrial plants with high emissions. The emission reductions in 2008 and 2009 were mainly due to the international financial crisis and the subsequent sharp decline in industrial production and freight transport. While the COVID-19 pandemic led to a decrease in travel in 2020 (particularly cruise traffic), the effect was partly offset by increased freight transport and a rapid return to normal levels thereafter.

The Norwegian Environment Agency often points out that large annual declines are due to temporary shutdowns at industrial facilities. Only in recent years do we see the decline in GHG emissions more systematically linked to increased climate policy ambitions. This results from the significant shift to electric vehicles and biofuels, as well as the electrification of oil and gas production.

However, a "methodological error" constitutes the most serious breach of the climate targets. Norwegian authorities, like those in many other countries, often draw a straight line between current emission levels and a future target when illustrating how developments must proceed in the coming years to reach the goal (see example in Fig. 7.2). This is akin to looking at the speedometer instead of the fuel gauge to determine whether we will make it home after a trip in our old fossil-fueled car. This is where Norwegian and other countries' authorities make the truly significant blunder.

---

[2] Ministry of Climate and Environment (2025).

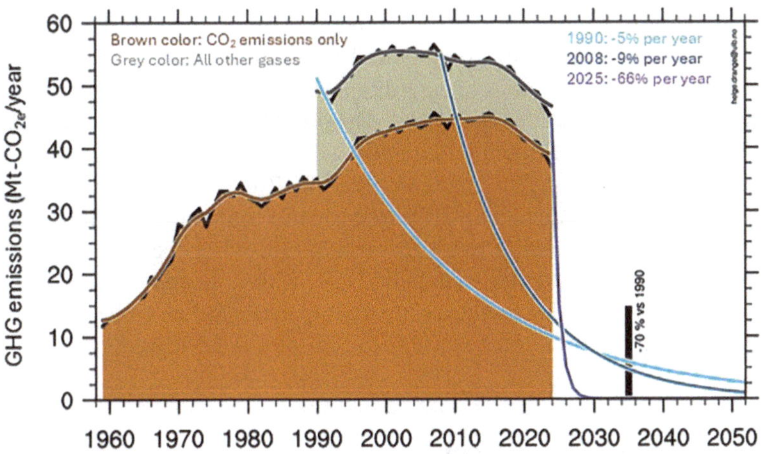

Fig. 7.3 Emission pathways that, with a 67% probability, are consistent with the 1.5 °C and 2 °C targets based on the Global Carbon Budget and assuming an equal percentage reduction for all countries[3]

I have often relied on analyses and figures produced by my esteemed colleague Helge Drange at the University of Bergen to illustrate how Norwegian GHG emissions compare to our climate targets and international climate commitments. He looks at the fuel gauge, not the speedometer. That is, he bases his approach on the logic inherent in the scientific concept of the carbon budget, captured in the popular saying "the journey is the destination."

What truly determines whether we achieve our climate goals is the sum of annual GHG emissions on the way to the target, not whether we manage to reach a specified emission level in a given target year—for example, 2030 or 2050. This logic presents a completely different and, in practice, more dramatic picture of the path toward the goal (see Fig. 7.3) than the more common (and scientifically incorrect) figures frequently used by the authorities (see Fig. 7.2).

When carbon budget logic is applied, the political and practical costs of delaying radical measures become glaringly apparent: accumulated emissions build up, creating an increasingly desperate need for drastic cuts if the target is to remain within reach. Figure 7.3 illustrates this relationship as the annual percentage reduction required for different starting years, given the actual emission levels at that time. The figure reveals the naked truth of how Norway has placed itself in an impossible position to deliver a globally fair contribution to the 1.5° target. If all countries were to contribute equally rapid percentage reductions, Norway would, in theory, have to reduce its emissions by 64% per year from 2026 onward.

Norway reached this impossible impasse first because it delayed adopting sufficiently ambitious climate targets (as seen in our first target, which aimed only to stabilize emissions), and second because we failed to reach prevailing but still modest

[3] Adapted by Helge Drange, based on: Lamboll et al. (2023).

goals. Had we applied the logic of carbon budgets and a globally fair distribution of reductions as early as 1990, the required annual reduction would have been a manageable 4%. Norway has achieved this level three times (in 1991, 2023, and 2024) and came close in three other years. The problem is that in all other years, the annual development was either too low or showed an actual increase—with 1998 as the record year showing an 8.0% increase (see Fig. 7.1).

Even if Norway had waited until 2008 to introduce a climate target based on these principles, the required annual reduction would still have been within a realistic range (8%). But as the situation stands in 2026, with a theoretical requirement of 64% annual reductions, other countries must compensate for the emission reductions that Norway cannot deliver if the world is to achieve the 1.5° target.

And Behind It Lurks the Oil

One of the partners in Noradapt, the Centre for Climate and Energy Transformation (CET) at the University of Bergen, has since 2015 organized a biennial conference entitled "Beyond Oil." The conference focuses on how Norway can navigate and move beyond the petroleum era. During the 2025 edition, Naomi Oreskes, author of the groundbreaking book *Merchants of Doubt*, made the following statement regarding the phase-out of Norwegian oil and gas production[4]:

> This is a tremendous opportunity to do the right thing. To use that money to finance a transition in an orderly way. And I mean, if any country could do it, it's Norway. And if Norway fails to do it, then that will be a colossal moral, economic, social, and historic failure. And Norway will go down in history as having absolutely failed at something they could have succeeded at.

So far, Norway has *not* seized this opportunity, and there are no political signals suggesting that such a move will occur soon. In Fig. 7.4, I have supplemented the introductory data on petroleum production with additional data illustrating exploration activity—specifically the allocation of exploration and production licenses. From the figure, we can discern two important points: First, there are no indications that Norway is in the process of scaling down petroleum production; rather, the opposite is true. Second, the intensity of exploration for new commercially viable discoveries has increased significantly in recent years.

The Norwegian Offshore Directorate (known as the Norwegian Petroleum Directorate until January 1, 2024) presented three scenarios in 2025—or "opportunity outlooks" as they prefer to call it—for petroleum extraction up to 2050: *high*, *base*, and *low*. Common to all these scenarios is the absence of any assumption regarding a politically managed phase-out. The primary distinction between them lies in the assumptions about when resources will be depleted. The critical uncertainty remains in the northern areas: will exploration in the Barents Sea reveal commercially viable discoveries? The *high* scenario suggests 2050 production levels roughly equal to

[4] Forchhammer (2025, December).

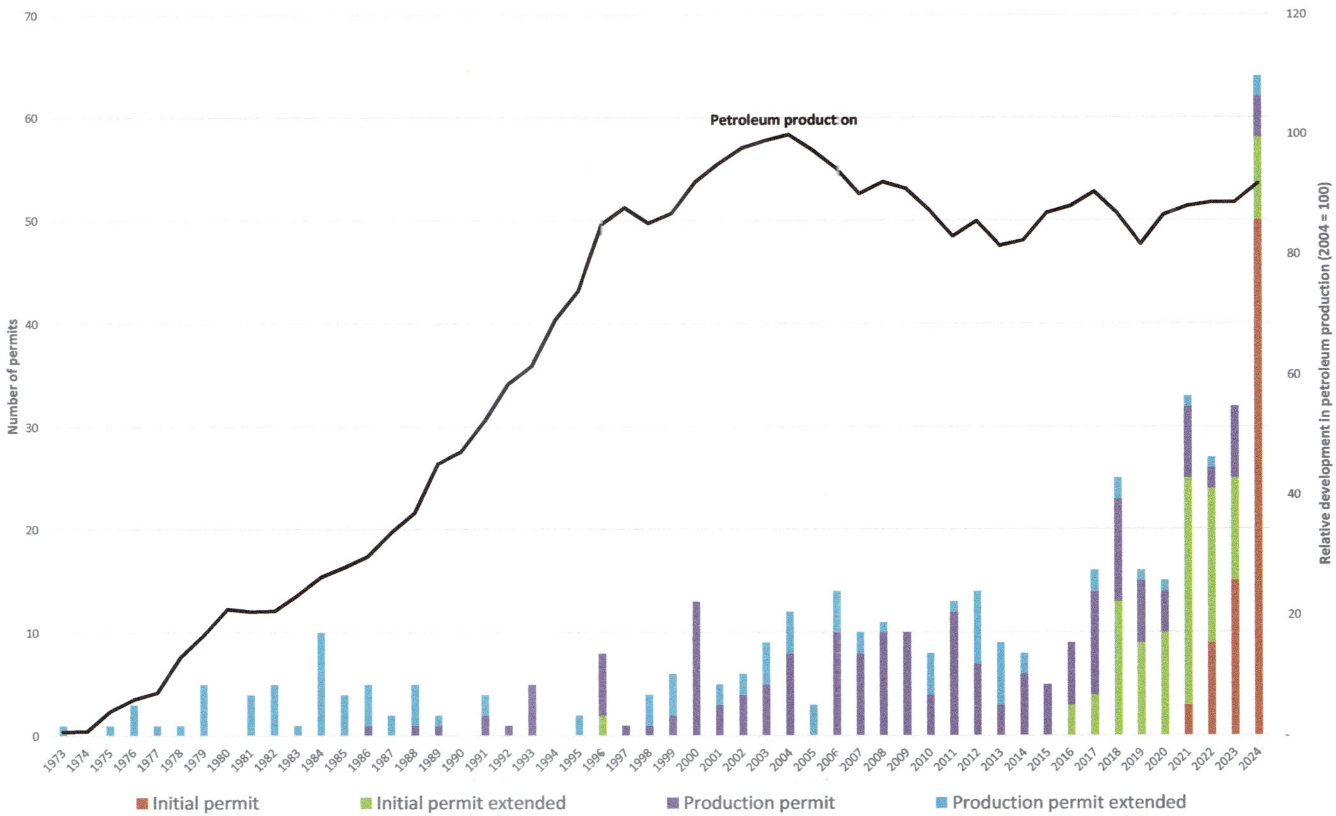

Fig. 7.4 Relative development of total petroleum production in Norway (2004 = 100) and annual numbers of exploration and production licenses awarded in the Barents Sea, North Sea, and Norwegian Sea[5]

[5] Norwegian Petroleum (2025).

those of 1990, while the *base* scenario sees a 34% reduction. Only the *low* scenario entails a nearly 100% reduction by 2050.

The most problematic aspect of the Directorate's analysis is the following assertion[6]:

> The three scenarios fall within the range of the global decline in oil and gas production estimated by the UN Intergovernmental Panel on Climate Change and the IEA to be consistent with successful implementation of the Paris Agreement.

A "counter-report" prepared for the Norwegian Climate Foundation by climate researchers at CICERO—including Glen Peters, one of the world's most cited climate scientists—points out that these scenarios are outdated. Commenting on the Directorate's conclusions, they noted[7]:

> Assessments based on global scenarios and actual developments in recent years *may* support the view that the *lowest* projected outcome is consistent with the Paris Agreement, but *not* the other two.

This represents a serious professional critique of a public body. It follows previous criticisms regarding methodological weaknesses, an excessive emphasis on industry gains at the expense of climate risk, and accusations of acting more as a political oil lobby than an administrative body with scientific integrity. The Office of the Auditor General has also repeatedly criticized the lack of coordination between environmental and petroleum management, describing the situation as reprehensible.[8]

Behind the assessments made by the Norwegian Offshore Directorate lies a broad political majority in the Norwegian Parliament supporting the continuation of the Norwegian petroleum era, encapsulated in a slogan formulated in 2019 by Jonas Gahr Støre, then Prime Minister from the Labour Party (*Arbeiderpartiet*) and former personal advisor to Gro Harlem Brundtland during her tenure as Director-General of the World Health Organization: "Norway should develop, not dismantle, the oil industry."

In a comment on his New Year's address and the presentation of the Labour Party's annual plan for Norway in 2026, Prime Minister Jonas Gahr Støre, who is still in office at the time of writing, made the following statement to Norway's largest daily newspaper[9]:

> My message today is clear... The oil and gas industry will continue to explore, produce, and supply Europe. In order to mitigate the decline in production on the Norwegian continental shelf, continued exploration activity is essential.

Consequently, within the framework of international climate policy—which targets emissions physically occurring within national borders—the central task of Norwegian climate policy becomes reducing emissions *from the production* of oil and gas. Since the mid-2010s, these production emissions have been the largest of Norway's

---

[6] Norwegian Offshore Directorate (2024).

[7] Peters and Andrew (2025).

[8] Office of the Auditor General (2018).

[9] Verdens Gang (2026, January 12).

eight main emission categories. As long as these specific emissions decrease—even though they represent only a tiny fraction of the emissions from the eventual use of the oil—the political majority sees no relevance in discussing reductions in exploration or production volumes.

What, then, will it cost in monetary terms, and not least in terms of "nature"—to achieve a goal of zero emissions from Norwegian petroleum extraction? An assessment by the Norwegian Climate Foundation shows that as of mid-2024, fourteen out of thirty-eight fields or field areas are fully or partially supplied with electricity from the power grid or from locally (at the platforms) produced offshore wind power.[10] This is ten more than in 2015. Thirteen fields or field areas are planned or considered for electrification in 2025. This also includes fields that are currently partially supplied with electricity but are not fully electrified.

In the 2024 edition of the annual resource reports from the Norwegian Offshore Directorate, it is estimated that electrification has reduced emissions from Norwegian oil and gas production by 3.39 million tons of CO_2 equivalents, corresponding to a reduction of about 30 compared to a scenario without electrification (see Footnote 6). However, emissions from Norwegian oil and gas production were still 40% higher than in 1990.

In 2024, Norwegian oil and gas production used 11.8 TWh of electricity (see Footnote 10). If we make a very conservative estimate, assuming the same amount of electricity is required to reduce current emissions from Norwegian oil and gas production as has been used for reductions so far, this implies a need for approximately 40 TWh of electricity. This is an absolute minimum estimate, given that the simplest and most cost-effective facilities—those requiring the least electricity—have been prioritized first. Even this minimum estimate means that full electrification of Norwegian oil and gas production would require at least an amount of electricity equivalent to about one quarter of Norway's total current electricity production. If there is to be room for both the electrification of the rest of society and the desired growth in new export industries that could replace something close to the current export value from oil and gas—as the majority of political parties in Norway are aiming for—then Norwegian climate policy going forward must focus on one thing: a massive expansion of new renewable energy. This recognition is also reflected in the title of the latest Norwegian government report on energy: "More of Everything—Faster."[11]

Shifting Crisis Response: From Moral Frugality to "Buying Our Way Out"

As I described in the introduction to this book, the 1974 oil crisis marked a watershed moment in Norwegian history. It was an era that taught us the necessity of *nøysomhet*—a traditional, highly esteemed Norwegian moral quality that can be translated as "frugality" or "contentment with little." We saw

[10] Statistics Norway (SSB) (2025).

[11] Norges offentlige utredninger (NOU) (2023).

this ethos vividly during the introduction of car-free Sundays and a collective encounter with the fear of empty fuel tanks. However, this approach has since undergone a profound transformation.

When the energy crisis struck in 2021/2022, Norway chose a path distinct from the rest of Europe by introducing the "Norway Price" (*Norges-pris*) for electricity—an automatic subsidy scheme that effectively placed a state-funded cap on household bills. While neighboring countries like Denmark and Sweden opted for targeted, short-term emergency measures, Norway nationalized the energy risk for private individuals on an apparently indefinite basis.

This pattern repeated itself when the conflict in Iran triggered the new global oil crisis in 2026. The price of petrol rose toward 30 NOK per liter. In constant 2026-NOK, that's in fact 100% higher than the price level during the first oil crisis in 1974. Large-scale protests followed. Transport industry activists threatened to block the main exit roads from major cities during the Easter holiday—the peak travel period in Norway, when "everyone" heads to their mountain cabins. So, the Norwegian Parliament (the Storting) enacted what was essentially a "Norway Price" for petrol and diesel by temporarily suspending the road use tax. For Norwegian households, this meant an immediate price reduction of up to 4.50 NOK per liter—a move that stands in sharp contrast to our European neighbors, who largely allowed high prices to function as a necessary incentive for structural transition.

Consequently, Norwegian consumers have ended up with a double-layered model of economic security: First, decades of climate policy and electric vehicle (EV) incentives have slashed nearly 60% of average household transport costs—compared to the technology available during the 1974 crisis. Then, when fossil fuels nevertheless become expensive for the remaining internal combustion vehicles, the state intervenes to remove the taxes on those fuels too, reducing the average cost for households by an additional 5%.

Perhaps it's as simple—and as sombre—as this: our oil wealth has cultivated the attitudes captured in the old Norwegian proverb, *"mye vil ha mer"* (literally: "much wants more," or "the more one has, the more one wants"). It evidently wasn't enough that Norway's electric vehicle transition has ensured that the purchasing power-adjusted costs for an average household to operate a private car are 60% lower during the second oil crisis compared to the first—even though the real pump price for petrol has doubled. With the "Norway Price" applied to both electricity and petrol, driving costs will be 65% lower during the 2026 crisis than they were in the 1974 version.

Nøysomhet—the once-venerated virtue of frugality—has clearly been abandoned as a crisis strategy for Norwegian households.

The Broad Norwegian Political Support for Continuing the Fossil Age

"All power shall be gathered in this hall" is a quote often used in Norwegian political debate to emphasize that the Storting should be the highest authority in Norway. The quote originates from a speech given by Member of Parliament Johan Sverdrup on March 8, 1872, in connection with a debate about ministers' access to the parliamentary chamber. It was linked to the desire to bring both members of parliament and ministers together under one roof. An obvious candidate for explaining the mismatch between scientifically justified climate policy ambitions and climate policy realities is therefore to take a closer look at the political power dynamics within the Storting itself.

In my youth—the early 1980s—I was active in what was then undoubtedly Norway's leading environmental party: the Liberal Party (*Venstre*). As early as 1988, the Green Party (*Miljøpartiet De Grønne* [MDG]) established itself as a political party. I was present at the founding meeting together with a small delegation of members from the Young Liberals (*Unge Venstre*). At that time, however, we considered the Liberal Party to be a better and more effective environmental political party. Most of us therefore chose to continue our engagement within the Liberal Party.

However, as the 2000s progressed, it became clear that more people were changing their minds about this—both politically active individuals and voters. In 2013, MDG achieved a national breakthrough with its first elected representative in the Storting. As a result of MDG's political success, the Socialist Left Party (*Sosialistisk Venstreparti* [SV]) strengthened its environmental profile throughout the 2010s. At the next parliamentary election—in 2017—the Red Party (*Rødt*) entered the Storting for the first time with one representative. Since its establishment in 2007, the Red Party has worked for "a society where freedom, economic justice, international solidarity, and concern for nature are central" (quote from the founding declaration in 2007). At the time of writing—autumn 2025—both the Red Party and MDG achieved another breakthrough by surpassing the Norwegian electoral threshold of 4%, resulting in an increase to nine and eight representatives, respectively. The Liberal Party did not reach the threshold, and with 3.7% of the votes, they secured only three representatives.

In today's Norwegian political landscape, four parties compete to be the best on climate and environment: the Liberal Party, MDG, SV, and the even more radical Red Party. These four parties consistently rank at the top and score similarly in assessments conducted by various environmental organizations ahead of municipal and parliamentary elections. Collectively, these parties have received around 20% of the vote in the last two parliamentary elections, in 2021 and 2025. Previously, since as far back as 1989, their share hovered around 15%, but overall—from the launch of the Brundtland Report and the goal of sustainable development—the parties that prioritize environmental issues have seen only a modest increase.

Is Hope Found in the Municipalities?

I have devoted considerable space to describing how so-called pioneering municipalities in Norway, ever since the World Commission's report introduced the sustainability goal in 1987, have served as a radical driving force in climate and sustainability efforts. These municipalities have been ahead not only of other municipalities, but also of national authorities. Assuming the role of a pioneering municipality has in many cases required strategic collaboration with radical environmental organizations (such as the Idébanken Foundation), progressively oriented business organizations (such as Hanen, which organizes small-scale farm food and tourism enterprises), progressively oriented interest organizations (such as the Norwegian Association of Local and Regional Authorities [KS]), progressively minded individuals within the state administration, progressively oriented national politicians (such as Torbjørn Berntsen during his tenure as Minister of the Environment for the Labour Party), and—not least—socially critical research communities (such as Western Norway Research Institute [WNRI]).

These structures still exist. But what about the political capacity for change in local elections? If we examine the municipal and county council elections, the share for the four environmental parties is consistently about 2 percentage points higher than in national elections. This is still low—far from a majority—but since the percentages for the municipalities conceal variations between municipalities, we need to delve deeper into the matter to uncover any potential frontrunner municipality effect.

In Fig. 7.5, I have constructed two indices to track the development of "green counterpower" from 1990 to 2025. At the national level, I have created an index consisting of two indicators: one for the proportion of parliamentary representatives from the four "green" parties, and another for the share these four parties have held in government during the same period. At the local level, I have constructed an index based on two indicators: the proportion of municipalities with the executive committee model where the four "green" parties have held the positions of mayor or deputy mayor, and—for those municipalities that have introduced parliamentarism (only the largest city municipalities)—the share of the four parties in the sitting city government at any given time. For both sets of indicators, I have summed the figures and then normalized the two indices by setting the value for 1990 to 100. I have assigned the same index value for all years within a given parliamentary or municipal election period.

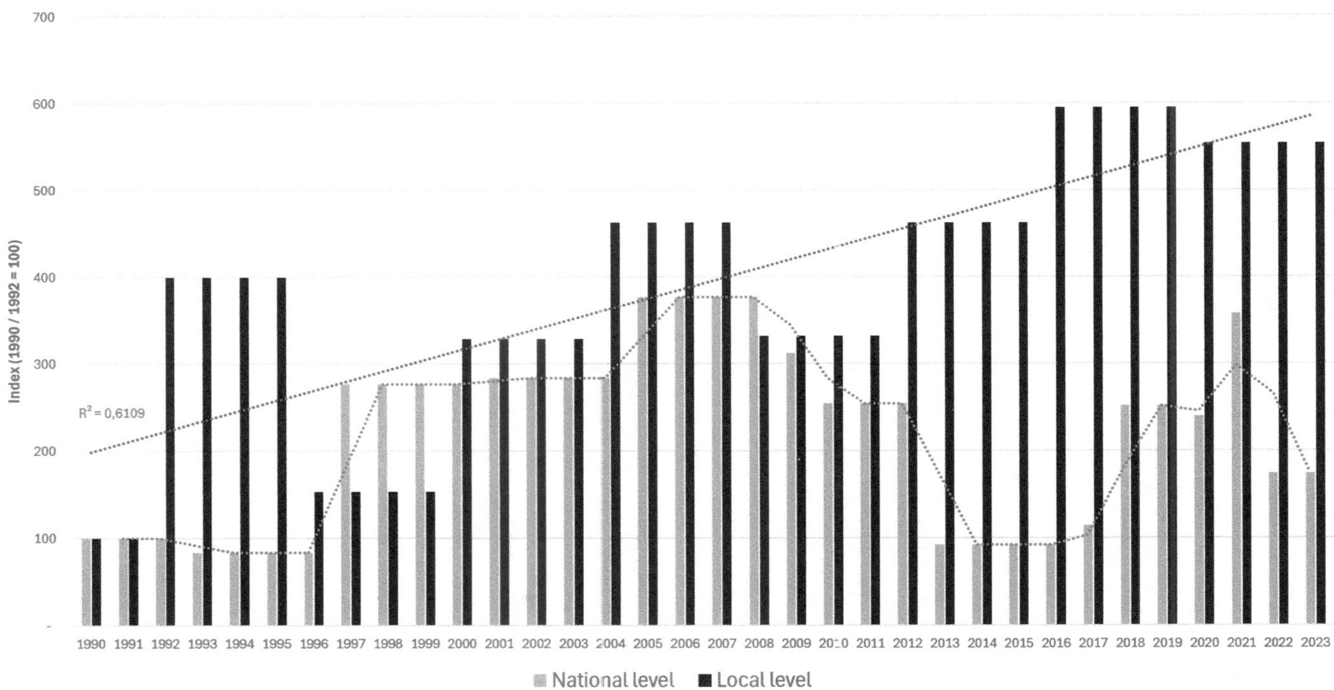

Fig. 7.5 Indicator for the development of political influence for the parties that the environmental movement in Norway ranks as most concerned with climate and sustainability (the Green Party, the Liberal Party, the Socialist Left Party, and the Red Party) at the national and local levels

The analysis illustrates what I described in the chapter on municipal environmental protection as a variant of political cleavage—an increasing environmental policy divide between the national and local levels in Norway. The analysis reveals a fundamental gap between two levels of administration. By using the four parties that the environmental movement consistently ranks as the most credible as indicators of a "green counterpower," we see the contours of a development in which radical environmental policy has become institutionalized in a few pioneering municipalities, while it has become fragmented and diluted at the national level.

If we limit ourselves to only the figures in the chart, the message is clear: the index for green municipal power (adjusted for parliamentarism and population weighting) follows a distinct linear trend line with an explanatory power (*R*-squared) of as high as 0.61. In stark contrast stands the index for green state power, where the linear explanatory power is almost non-existent. The moving average shown in the figure illustrates how the green counterpower fluctuated during the same period, peaking around the turn of the millennium.

The decline of green counterpower at the national level cannot be understood without considering the prevailing orientations within the major governing parties—particularly the Labour Party. The period before the turn of the millennium was marked by a genuine ideological confrontation. With figures such as Torbjørn Berntsen as Minister of the Environment, we saw a Labour Party that at least dared to challenge the boundaries of material growth and speak of absolute ecological limits. This was a period when state power flirted with the radical implications of the sustainability goal.

Since then, we have witnessed a dramatic transformation. Today's Labour Party has largely abandoned this radical self-awareness in favor of a policy characterized by slogans such as "Norway shall develop, not dismantle, the oil industry"—where the framework conditions of the petroleum sector and economic growth take precedence over ecological considerations. This shift in green governmental power explains why the national index is so volatile: when the green parties have participated in social democratic or conservative governments, they have encountered resistance not only from traditionally less environmentally engaged parties like the Centre Party (*Senterpartiet*) and the Progress Party (*Fremskrittspartiet*), but also from a social democracy that has redefined sustainability to mean "green growth" within existing frameworks.

But Was the Green Party (MDG) So Green After All?

In 2017—five years after the IPCC first attempted to define transformation in a climate context—the Research Council of Norway hosted its inaugural conference on the topic, titled: "Transition 2017: To think it, to wish it, to want it—but to do it?" Lan Marie Berg, then Deputy Mayor of Oslo and City Commissioner for Environment and Transport from MDG, opened the conference. Speaking on Oslo's climate efforts, she began with a pointed statement: "We must reduce car traffic in Oslo so that the city can continue to grow."

At the time, I had recently joined MDG and was part of an internal digital forum for Volunteers and Resource Persons. As I listened to her speech, I posted a simple question to the group: "Why is it important for MDG that Oslo should continue to grow?"

Within a few hours, I received 173 comments. Only 5% of the respondents shared my skepticism toward continued urban growth. The remaining 95% supported the growth mandate. Their arguments felt surprisingly reminiscent of the traditional growth-oriented parties.

A few days later, I was invited to speak at the MDG county convention in Sogn og Fjordane. To an audience gathered far from the capital, I presented a summary of the arguments from the "growth-wing" of the party, paired with my own counterarguments. It was met with spontaneous applause.

Arguments from the MDG "Growth Wing" presented in the MDG internal digital forum.	My responses presented to the MDG County Meeting.
We cannot turn back; we must simply adapt to developments.	Is the Green Party not a system-critical party, after all?
Growth in large cities is climate-friendly.	While local growth within a city might appear green, it often creates negative impacts elsewhere.
Urban residents travel less than rural residents.	Everyday travel, "yes;" but when taking into account leisure travel—in particular by plane—definitely "no."
Emissions per capita are lower in cities than in rural areas.	Statistics show the opposite when consumption is included.
People are moving from rural areas to cities.	Urban growth is primarily driven by immigration; major cities see a net outmigration of Norwegian-born residents.
It is impossible to relocate government jobs from Oslo.	It creates political noise, certainly, but it is entirely possible—as proven several times in Sogn og Fjordane.

How to Get Norwegian Environmental Policy Back on a Sustainable Track

In 2010, Karl Georg Høyer published an important article entitled "Seven Theses on CO_2-Reductionism and its Interdisciplinary Counteraction."[12] At that time, he had moved to Oslo and begun working as a professor of technology, design, and environment at Oslo University College, while still holding a 20% position at WNRI. Høyer criticizes the prevailing scientific and political environmental discourse for being dominated by what he termed *CO_2-reductionism*, where complex socio-ecological phenomena are reduced to the single issue of carbon emissions. He argued that this reductionist perspective was dangerous, as neither the climate crisis nor the broader sustainability crisis can be solved through such a narrow lens. He warned that environmental measures designed within this silo risked creating new, equally serious problems. To address this, he presented seven theses:

1. Reconcile CO_2 with other GHGs.
2. Reconcile CO_2 with fossil energy.
3. Reconcile CO_2 with energy.
4. Reconcile CO_2 with consumption.
5. Reconcile CO_2 with economic growth.
6. Reconcile CO_2 with sustainable development.
7. Integrate CO_2 with low-carbon society.

Høyer's critique was primarily aimed at national and supranational policy. However, even if we recognize the emergence of a municipal "green counterforce," local environmental policy has also fallen prey to its own form of reductionism. Inspired by Høyer, I will conclude by formulating seven equivalent theses on how to address the reductionism that has permeated Norwegian municipalities. Like Høyer's, my theses are presented as steps on a ladder, moving from specific technical reintegration to comprehensive systemic transformation.

1. Reintegrate local environmental policy with local pollution policy.
2. Reintegrate local environmental policy with energy policy.
3. Reintegrate local environmental policy with consumption.
4. Reintegrate local environmental policy with nature conservation policy.
5. Reintegrate local environmental policy with sustainable development.
6. Align environmental policy with the discourse on degrowth.
7. Revitalize cooperation between pioneering municipalities and radical transformation forces in national policy.

Although I have emphasized the responsibility of municipalities to address global issues, this does not mean that conventional *local pollution* control is unimportant. Today, local pollution efforts are squeezed from two sides: they are often deprioritized within an overly broad sustainability debate, or they vanish entirely within a

[12] Høyer (2010).

narrow focus on GHG emissions. In many cases, a robust pollution policy offers clear synergies with climate action. For instance, the runoff of nutrients (phosphorus and nitrogen) from municipal sewage exacerbates the acidification of coastal waters already stressed by atmospheric CO_2 concentrations. This link makes nutrient removal even more critical than local considerations alone would suggest.[13]

Regarding *energy*, a divide occurred in the 2000s when the state excluded electricity-saving measures from national climate policy. The rationale was that since Norwegian hydropower is defined as emission-free, saving electricity does not count toward UN GHG emission reporting. While the early pilot municipalities highlighted energy efficiency as a cornerstone of climate policy,[14] subsequent state support schemes effectively severed the link between energy and climate at the local level.

In the chapter on sustainable *consumption*, I described Torbjørn Berntsen's radical initiatives from the mid-1990s to reduce consumption in wealthy countries. While this led to local experiments under Local Agenda 21, these eventually lost momentum and failed to shift national policy toward reducing Norwegians' material footprint. While municipalities have continued to experiment with addressing indirect emissions (emissions triggered by Norwegian consumption but occurring abroad), these efforts have largely been bottom-up initiatives without national backing.

The urgency of reintegrating *nature conservation* was brought to the forefront of public debate in early 2024 by the NRK series *Oppsynsmannen* (*The Ranger*), featuring comedian Bård Tufte Johansen. The series highlighted how the bit-by-bit encroachment on natural areas—for cabins, roads, and commercial zones—accumulates into an ecological crisis. Combined with data-driven maps from the Norwegian Institute for Nature Research, which showed that Norway lost 79 square meters of nature per minute between 2017 and 2022, this sparked a revival of the 1970s conservation debate, reframed as a modern local biodiversity crisis.[15]

My fifth thesis is a two-way challenge. *Local climate policy* must respect broader *sustainability* boundaries, but municipal sustainability work must also be narrowed down to the issues that are truly critical. A lack of sustainability framing was evident in the dramatic halt of onshore wind power. After a surge in development up to 2020, intense local opposition led the government to grant municipalities a virtual veto over new projects. Annual expansion plummeted from 220 MW (2000–2022) to just 4 MW in 2025.[16] While this move strengthened local participation in the spirit of Local Agenda 21, it illustrates how climate policy can stall when disconnected from broader local values.

The sixth thesis leads to the importance of aligning local policy with the discourse on *degrowth*. While increasing living standards is vital for low-income countries, claiming that continued economic growth in one of the world's wealthiest nations

---

[13] Dannevig et al. (2019).

[14] Groven and Aall (2002).

[15] Norsk institutt for naturforskning (NINA) & Norsk rikskringkasting (NRK) (2024, January 10).

[16] Norges vassdrags- og energidirektorat (NVE) (2024).

is a legitimate sustainability goal is a fallacy. For Norway, economic degrowth—managed in a controlled and fair manner—is a prerequisite for true sustainability. This has been a central tenet of critical Norwegian environmental research since the 1960s and has recently resurfaced through organizations like Rethinking Economics Norway.

My *final* thesis concerns the restoration of the alliances prevalent in the 1990s between radical national forces and pioneering municipalities. Programs like the MIK initiative proved that change is possible when local niches are protected and supported. This is what transition theory calls *strategic niche management*: the process where radical innovations emerge in protected environments before challenging and eventually transforming the established system. Research confirms that when the state uses such niches to promote sustainable innovation, it can be highly effective.[17]

What, Then, Can the World Learn from "Looking to Norway"?

I began this book by highlighting the favorable conditions Norway possessed for meeting the formidable challenge of a sustainable climate transition when the process was officially launched following the Brundtland Report in 1987. These conditions still largely apply:

- *A well-functioning democracy*: Despite being subject to global pressures such as political extremism, a rise in corruption, and the polarization of public debate in the wake of social media, Norway's democratic foundation remains solid.
- *A robust local democracy*: Norway maintains a high degree of political and, to some extent, economic autonomy for its municipalities. Public support for and involvement in local democratic processes remains strong, even if it has seen a slight decline in recent decades.
- *Broad legitimacy for policy governance*: Authorities enjoy strong support from both the public and the private sector for using both "sticks" and "carrots." While there has been a gradual shift away from "sticks" (regulations and taxes) toward "carrots" (financial incentives and subsidies), the mandate for state intervention remains.
- *A society of trust*: Norway continues to be a high-trust society where the public and the business sector maintain confidence in authorities and scientific research—notwithstanding a relatively high proportion of climate skepticism compared to some other nations.
- *Economic capacity*: Due in large part to petroleum revenues, Norway possesses the financial strength to implement the costly measures required for a climate transition.

---

[17] Schot and Geels (2008).

Perhaps most importantly, for better or worse, is Norway's unique natural situation. As former Minister of Social Affairs Olav Bruvik famously stated at the Labour Party's national convention in 1961:

> It is fair to say that the Lord was in a good mood the day He created Norway.

At the time, this optimism reflected a view of nature as an inexhaustible source of welfare and industrial development. However, developments since the 1960s have shown that this very gift is now under threat. Norway's natural resources place us in a paradoxical position: we possess vast renewable energy resources—primarily hydropower and wind—yet we also hold the very resources that constitute the root cause of the climate problem. Our continued heavy investment in petroleum exploration means that Norway stands to benefit economically regardless of the direction of international climate policy.

Given this unique position, what can Norway and the rest of the world learn from the history I have outlined? Is there still reason to "look to Norway" in the global effort toward a sustainable transition? I would summarize the lessons in five key points:

- *Counter-expertise is necessary, but not sufficient*: Independent knowledge institutions, like the Western Norway Research Institute, are vital. Genuine independence requires economic and institutional autonomy—being a self-owning foundation that does not generate profit—and often, a limited size. "Small is good" applies here; we must maintain the framework conditions that allow such independent voices to exist.
- *Institutions are necessary, but not sufficient*: You can have the world's first Ministry of the Environment and the best experts, but if they lack the mandate to override economic drivers, transformation will not occur. The political struggle must focus on establishing institutions with the authority to demand and, if necessary, sanction insufficient efforts from other public bodies.
- *Technological innovations are important, but social innovations are more important—and the willingness to degrow is the most important*: Norway has "burned itself" by abandoning ambitions for radical social change in favor of costly technological fixes. Our petroleum wealth has made it possible to bypass difficult political discussions about capping and eventually reducing our material standard of living.
- *Do not subsume all good causes under the sustainability label—sustainable development is "something," not "everything"*: Norway was a pioneer in attempting to elevate sustainability to the same level as economic considerations. However, these ambitions were watered down as more political concerns—including conventional economic interests—were added to the label. This has emptied the sustainability goal of its specific political content and steering power.
- *Symbolic politics is not necessarily wasted*: While critics dismiss symbolic politics as a distraction, symbols help shape reality. The long and tiring debate over what sustainable tourism means eventually led the Norwegian tourism sector to accept that travel-related emissions must be addressed rather than ignored.

As I finish writing this in early spring 2026, the world faces a transformative moment. Many fear that internationally binding global cooperation is disintegrating in favor of regionalism and superpower unilateralism. If major countries stop participating, regional climate targets become almost meaningless.

Where, then, does my hope lie? I look back to the history of Local Agenda 21 (LA21). This initiative emerged from a bench proposal during the Rio Conference in 1992, driven by representatives from local authorities who succeeded in including a dedicated chapter on their role. Subsequent evaluations showed that the greatest successes of Agenda 21 were observed at the local level.

It may seem naive to believe that global change can be driven by local initiatives. However, the most important reason for advancing LA21 was the recognition of democracy as a necessary precondition for sustainability. While the UN cannot explicitly mandate liberal democracy to member states, it can support LA21 processes as a bottom-up driver of democratization.

Considering current political unrest and the fraying of global cooperation, the local political level can serve as a safe haven—a genuine countervailing force that keeps the struggle for a sustainable climate transition alive while we await the return of global cooperation. In this situation, it is more important than ever to seek out radical partners and build strategic niche alliances between the local and national levels.

References

Roosevelt, F. D. (1942, September 16). *Remarks on the transfer of a naval vessel to Norway.* The American Presidency Project. https://www.presidency.ucsb.edu/node/209930

Ministry of Climate and Environment. (2025). *Regjeringens klimastatus og -plan for 2026* [The Government's climate status and plan for 2026] (Prop. 1 S [2025–2026]). https://www.regjeringen.no/contentassets/2f9d09306aab43d4b277a8952beeacd3/no/pdfs/regjeringens-klimastatus-og-plan-for-2026.pdf

Lamboll, R. D., Nicholls, Z. R. J., Smith, C. J., Kikstra, J. S., Byers, E., & Rogelj, J. (2023). Assessing the size and uncertainty of remaining carbon budgets. *Nature Climate Change, 13*, 1360–1367. https://doi.org/10.1038/s41558-023-01848-5

Forchhammer, J. (2025, December). *Harvard professor Naomi Oreskes about Norway.* LinkedIn. https://www.linkedin.com/posts/julie-forchhammer-87999921b_harvard-professor-naomi-oreskes-about-norway-activity-7391384851003863040-IX_G/

Norwegian Petroleum. (2025). *Facts about Norwegian petroleum.* https://www.norskpetroleum.no/fakta/

Norwegian Offshore Directorate. (2024). *Ressursrapport 2024: Mulighetsbilder* [Resource Report 2024: Opportunity Outlooks]. https://www.sodir.no/aktuelt/publikasjoner/rapporter/ressursrapporter/ressursrapport-2024/mulighetsbilder/

Peters, G., & Andrew, R. (2025). *Paris-vennlig olje?* [Paris-friendly oil?]. Norwegian Climate Foundation. https://api.klimastiftelsen.no/wp-content/uploads/2025/12/Paris-vennlig-olje.pdf

Office of the Auditor General. (2018). *The Office of the Auditor General's investigation of the authorities' efforts to safegu2ard the environment and fisheries in petroleum activities in the northern areas.* Document 3:9 (2018–2019).

Verdens Gang. (2026, January 12). Statement by Prime Minister Jonas Gahr Støre on exploration activity. https://www.vg.no

Statistics Norway (SSB). *(2025). Table 11561: Greenhouse gas emissions from Norwegian economic activity.* https://www.ssb.no/statbank/table/11561

Norges offentlige utredninger (NOU). (2023). *Mer av alt—raskere* [More of everything—faster] (NOU 2023:3). Ministry of Petroleum and Energy. https://www.regjeringen.no/contentassets/5f15fcecae3143d1bf9cade7da6afe6e/no/pdfs/nou202320230003000dddpdfs.pdf

Høyer, K. G. (2010). Seven theses on CO2-reductionism and its interdisciplinary counteraction. In R. Bhaskar, C. Frank, K. G. Høyer, P. Næss & J. Parker, (Eds.), *Interdisciplinarity and climate change. Transforming knowledge and practice for our global future* (pp. 35–53). Routledge.

Dannevig, H., Mohus, T., Groven, K., Hovelsrud, G. K., Lundberg, A. K., Sandersen, H., Wallhead, P., Bellerby, R., & Johansen, A. (2019). *Forslag til politikk for forvaltning av havforsuring i norske kystfarvann.* [Proposed policy for managing ocean acidification in Norwegian coastal waters] (VF-report 6/2019). Western Norway Research Institute.

Groven, K., & Aall, C. (2002). *Lokal klima- og energiplanlegging. Norske kommunar som aktørar i klimapolitikken?* [Local climate and energy planning] (VF-report 12/02). Western Norway Research Institute.

Norsk institutt for naturforskning (NINA) & Norsk rikskringkasting (NRK). (2024, January 10). *Norge i rødt, hvitt og grått—en kartbasert dokumentasjon av naturtap* [Norway in red, white, and grey—a map-based documentation of nature loss]. NRK. https://www.nrk.no/dokumentar/xl/nrk-avslorer-hvordan-norges-100-storste-naturinngrep-ble-til_-naturen-vi-ofret-med-oynene-lukket-1.17168283

Norges vassdrags- og energidirektorat (NVE). (2024). *Data for commissioned onshore wind farms in Norway.* https://www.nve.no/energi/energisystem/vindkraft-paa-land/data-for-utbygde-vindkraftverk-i-norge/

Schot, J., & Geels, F. W. (2008). Strategic niche management and sustainable innovation journeys: Theory, findings, policy relevance, and critical reflections. *Technology Analysis & Strategic Management*, *20*(5), 537–554. https://doi.org/10.1080/09537320802292651

FSC
www.fsc.org
MIX
Papier | Fördert gute Waldnutzung
FSC® C083411

Kolibri 360 GmbH
Ferdinand-Jühlke-Straße 7
99095 Erfurt, Deutschland
produktsicherheit@kolibri360.de